Semantic Web Personalization and Context Awareness:

Management of Personal Identities and Social Networking

Militiadis Lytras
The American College of Greece, Greece

Patricia Ordóñez de Pablos
University of Oviedo, Spain

Ernesto Damiani
University of Milan, Italy

Senior Editorial Director:	Kristin Klinger
Director of Book Publications:	Julia Mosemann
Editorial Director:	Lindsay Johnston
Acquisitions Editor:	Erika Carter
Development Editor:	Joel Gamon
Production Coordinator:	Jamie Snavely
Typesetters:	Jennifer Romanchak and Michael Brehm
Cover Design:	Nick Newcomer

Published in the United States of America by
Information Science Reference (an imprint of IGI Global)
701 E. Chocolate Avenue
Hershey PA 17033
Tel: 717-533-8845
Fax: 717-533-8661
E-mail: cust@igi-global.com
Web site: http://www.igi-global.com/reference

Library of Congress Cataloging-in-Publication Data

Semantic Web personalization and context awareness: management of personal identities and social networking / Miltiadis Lytras, Patricia Ordonez de Pablos and Ernesto Damiani, editors.
p. cm.
Includes bibliographical references and index.
Summary: "This book communicates relevant recent research in Semantic Web-based personalization as applied to the context of information systems"-- Provided by publisher.
ISBN 978-1-61520-921-7 (hardcover) -- ISBN 978-1-61520-922-4 (ebook) 1. Semantic Web. 2. Context-aware computing. 3. Identification. 4. Online social networks. I. Lytras, Miltiadis D., 1973- II. Ordonez de Pablos, Patricia, 1975- III. Damiani, Ernesto, 1960-
TK5105.88815.S436 2011
025.042'7--dc22
2010054441

British Cataloguing in Publication Data
A Cataloguing in Publication record for this book is available from the British Library.

To my parents, Elvira and Joaquín

—Patricia

Editorial Advisory Board

List of Reviewers

Antonio Zilli, *University of Salento, Italy*
B. Cristina Pelayo García-Bustelo, *University of Oviedo, Spain*
C. Bisconti, *University of Salento, Italy*
Eduardo Peis, *University of Granada, Spain*
Eliana Campi, *University of Salento, Italy*
Enrique Herrera-Viedma, *University of Granada, Spain*
F. Grippa, *University of Salento, Italy*
Giuseppina Passiante, *University of Salento, Italy*
Jairo Francisco de Souza, *Federal University of Juiz de Fora, Brazil*
Javier Criado, *University of Almería, Spain*
Jordán Pascual Espada, *University of Oviedo, Spain*
Jordi Conesa, *Universitat Oberta de Catalunya, Spain*
José Emilio Labra-Gayo, *Universidad de Oviedo, Spain*
José M. Morales-del-Castillo, *University of Granada, Spain*
José-Andrés Asensio, *University of Almería, Spain*
Juan Miguel Gómez Berbís, *Universidad Carlos III de Madrid, Spain*
Kerkiri Al. Tania, *University of Macedonia, Greece*
Konetas Dimitris, *University of Ioannina, Greece*
Avgoustos. A. Tsinakos, *University of Kavala Institute of Technology, Greece*
Luis Iribarne, *University of Almería, Spain*
M. De Maggio, *University of Salento, Italy*
Margaret Fitzgerald-Sisk, *University of Minnesota, USA*
Nicolás Padilla *University of Almería, Spain*
Nouha Taifi, *University of Salento, Italy*
Oscar Sanjuán Martínez, *University of Oviedo, Spain*
Ricardo Colomo-Palacios, *Universidad Carlos III de Madrid, Spain*
Rubens Nascimento Melo, *Pontifical Catholic University of Rio de Janeiro, Brazil*
S. Totaro, *University of Salento, Italy*
Santi Caballé, *Universitat Oberta de Catalunya, Spain*
Sean Wolfgand Matsui Siqueira, *Federal University of the State of Rio de Janeiro, Brazil*
Valerio Cisternino, *University of Salento, Italy*
Vladlena Benson, *Kingston University, UK*

Table of Contents

Detailed Table of Contents

The process of summarizing information is becoming increasingly important in the light of recent advances in resource creation and distribution and the resulting influx of large numbers of information in everyday life. These advances are also challenging educational institutions to adopt the opportunities of distributed knowledge sharing and communication. Among the most recent trends, the availability of social communication networks, knowledge representation and of activate learning gives rise for a new landscape of learning as a networked, situated, contextual and life-long activities. In this scenario, new perspectives on learning and teaching processes must be developed and supported, relating learning models, content-based tools, social organization and knowledge sharing.

Based on the literature review of the theory of trust, this chapter aims to provide an insight into trust formation on social networking sites (SNS). An overview of the current state of cybercrime and known ways of threat mitigation helps shed some light on the reasons why social networks became easy targets for Internet criminals. Increasingly personalisation is seen as a method for counteracting attacks perpetrated via phishing messages. This paper aims to look specifically at trust in online social networks and how it influences vulnerability of users towards cybercrime. The chapter poses a question whether personalisation is the silver bullet to combat cyber threats on social networks. Further research directions are discussed.

This research aims to apply models extracted from the many-body quantum mechanics to describe social dynamics. It is intended to draw macroscopic characteristics of organizational communities starting from the analysis of microscopic interactions with respect to the node model. In this paper we intend to give an answer to the following question: which models of the quantum physics are suitable to represent the behaviour and the evolution of business processes? The innovative aspects of the project are related to the application of models and methods of the quantum mechanics to social systems. In order to validate the proposed mathematical model, we intend to define an open-source platform able to model nodes and interactions within a network, to visualize the macroscopic results through a digital representation of the social networks.

chapter argues through a conceptual model the strategic role of the integration of knowledge management systems and special communities for the acceleration of the new product development process and present an ontology-based knowledge management system and its application in the context of a community of automotive designers. More precisely, the issue management, based on this engineered IT-system, will accelerate and optimize the product design phase and knowledge sharing among the designers and engineers.

The advent of the information age represents both a challenge and an opportunity for Medicine. New forms of diagnosis, innovation-oriented supervision and expert location paths are deeply impacting medical sciences as we know it around the word. In this new scenario, Semantic Technologies can be seen as new and promising tool to support knowledge-based services, and particularly for the health domain, medical diagnosis. This paper presents MedFinder, a system based on Semantic Technologies and Social Web to improve patient care for Medical Diagnosis. The main breakthroughs of MedFinder are the follow-up once the diagnosis is performed, by using a medical ontology and formal reasoning together with rules, since it makes possible to locate the most appropriate doctor for a patient using Geographical Information Systems (GIS) and taking into account user preferences given via Social Web feedback.

Over the last years a great deal of ontologies of many different kinds and describing different domains has been created, and new methods and prototypes have been developed to search them easily. These ontologies may be reused for several tasks, such as for increasing semantic interoperability, improving searching, supporting information systems or the creation of their conceptual schemas. Searching an ontology that is relevant to the users' purpose is a big challenge, and when the user is able to find it a new challenge arises: how to adapt the ontology in order to be applied effectively into the problem domain. It is nearly impossible to find an ontology that can be applied as is to a particular problem. Is in that context where ontology refactoring takes special interest. This chapter tries to clarify what ontology refactoring is and presents a possible catalog of ontology refactoring operations.

Web-ontologies are becoming the de facto standard for WWW-based knowledge representation. As a consequence, user modeling has been associated to Web-ontologies. However, data schemes evolve, and therefore ontologies also evolve. Thus, adaptive systems, more than other ontology-based system, are directly affected by changes in ontologies. Because of this, it is important that adaptive systems can be prepared to deal with the problems that occur after changes are applied to ontologies. In this chapter, we perform a literature review on the field of ontology evolution aiming at serving as a point of reference for user modeling area. Therefore, adaptive systems developed on ontology-based user modeling could adapt to changes when the ontologies change.

The interactive tools like blogs, wikies etc. known under the commonly acceptable name Web2.0, led to a new generation of internet services and applications such as social networks, recommendation systems, reputation systems, etc, allowing for public participation in the formation of the content of the web, and at the same time fueling an explosion of information. This information is a widely available intellectual capital. Due to the opportunities that arise from the exploitation of this information, in this article we i) present the rationale under which these systems function, ii) summarize and apply in an indicative manner the mathematical models used to handle this information, iii) propose a general architecture of these systems and iv) describe a hybrid multifaceted algorithm that exploits the capabilities arising from this information towards a personalized inference for a specific user. The result of this work is an indication of the capabilities that arise from further exploitation of these systems.

The current chapter is a review of the variety of Student Models that have been reported in the literature. The chapter can virtually be divided in two parts: The first part outlines a number of typical examples of Student Models that have been developed in order to indicate why these models have been developed, what are their uses and the achievements of Student Models in Education. The second part discusses how the Student Models can be useful in Distance Education, what are the criteria for testing the applicability of such models and finally reports Student Models that may apply in Asynchronous Distance Education

Preface

As semantic web technologies prove their value with targeted applications, there are increasing opportunities to consider their application in social contexts for learning (at individual, group and organizational levels) and human and social development. Knowledge management has been accepted as critical to increasing knowledge-related performance through the more effective use of intellectual capital. Additionally, governments are forced to increasingly deal with knowledge services that form larger parts of the global economy and society. Thus, there are many examples of applications of semantics for empowering knowledge management or better supporting knowledge services for social networks.

The book also aims to communicate and disseminate recent research and success stories that show the power of semantics to improve upon traditional knowledge management and organizational learning approaches, or realize emerging requirements of knowledge services for social networks towards increased personalization.

The book has a clear editing strategy.

- To be the reference edition for all those interested in the applied aspects of semantic web based personalization in real world contexts
- To be the reference edition for all those (policy makers, government officers, academics and practitioners) interested in exploiting emerging semantics within knowledge intensive organizations for personalized solutions

Target Audience

The audience for this book includes:

- Politicians
- Professors in academia
- Policy Makers
- Government officers
- Students
- Corporate heads of firms
- Senior general managers
- Managing directors
- Board directors
- Academics and researchers in the field both in universities and business schools

- Information technology directors and managers
- Quality managers and directors
- Human resource directors
- Libraries and information centres serving the needs of the above

The unique characteristic of the proposed edition is that goes beyond the verbalism of wishful thinking and applies modern approaches through emerging technologies like:

- Reasoning and rules for user modeling and personalization in the Semantic Web
- Information extraction for capturing semantics
- Acquisition and application of user profiles
- Ontology aligning and evolution for user modeling
- User-aware querying in the Semantic Web
- Trust, privacy and transparency in the personalization process
- Semantic Web Services and service-oriented architectures for personalization
- Semantically enriched user models for ubiquitous and mobile environments
- Content-based filtering exploiting content semantics
- Social filtering (collaborative filtering) exploiting content semantics
- Hybrid filtering technologies using content semantics
- Filtering technologies for heterogeneous content based on semantics
- How to use semantics for preference description
- Needs: scenarios where personalization brings benefits
- Experiences: lessons learnt from previous research
- Solutions: for personalized access to content in a Semantic Web

Patricia Ordóñez de Pablos
Gijón (Spain), June 2010

Miltiadis D. Lytras
Attica (Greece), June 2010

Chapter 1
Transitioning a Face-To-Face Class to an Online Class: A Knowledge Engineering Narrative

Margaret Fitzgerald-Sisk
University of Minnesota, USA

Robert D. Tennyson
University of Minnesota, USA

ABSTRACT

This chapter speaks to knowledge engineering through the use of narrative, personal research. It is intended that the readers experience the development of an online class from the perspective of the instructional designer and the perspective of the subject matter expert and instructor for whom the instructional designer is working. In this experience, the reader will gain a view of knowledge engineering with respect to a number of issues and requirements regarding how to represent, create, manage and use ontologism as shared knowledge representations.

TRANSITIONING A FACE-TO-FACE CLASS TO AN ONLINE CLASS: A KNOWLEDGE ENGINEERING NARRATIVE

One of the most specific versions of knowledge engineering for organizational applications can be seen in the use of learning management systems (LMS) and learning content management systems (LCMS) in both business and academia (Lytras, Tennyson, & Ordonez de Pablos, 2009). Most institutions of higher education use some sort of LMS or LMS/LCMS combination to manage their online learning, as do most large businesses. At the University of Minnesota, the preferred tool is Moodle. It functions as both an LMS and an LCMS, the former in its interface with PeopleSoft to manage class enrollment and access to online classes, and the latter in its ability both to house content for classes and allow access to that content in whatever manner the instructor or instructional designer deems appropriate.

This chapter speaks to knowledge engineering through the use of narrative, personal research. It is our intention to allow the readers to experience the development of an online class from

DOI: 10.4018/978-1-61520-921-7.ch001

the perspective of the instructional designer and the perspective of the subject matter expert and instructor for whom the instructional designer was working.

Taking a Class to Design a Class

It all began the day the instructional designer decided she needed a more well-rounded understanding of what she was learning in pursuing the doctoral degree. She had learned most of what she thought she needed from all angles except the psychology angle. In the discovery process for classes she could take to round out that angle of understanding, she discovered one that was offered on the psychology of instruction and technology, which seemed perfect. Her background is in teaching, and her then-current field was information technology. She was deeply involved in several technology projects, including one instructional technology project, at work, and so it seemed a perfect fit. She was, at the time, a bit concerned that perhaps she had too much background in technology for this class, but she thought that the psychology angle would be the one she would get the most from. Little did she know how correct that thought was!

From the first class meeting, it seemed like she had a connection with the professor. He had a quirky sense of humor that spoke to her, especially when he asked her what the heck she was doing taking his class. She trotted out her explanation of the hole in her education, which seemed to satisfy. She kept her mouth shut about the 20-plus years of instructional design practice and training she had under her belt. That was an exercise in futility, though, because it is something one just cannot hide from an expert. It turned out that the class would be working together to understand technology and instruction in the guise of exploring how this instructor-led, in-person class could become a more student-led, online class.

Assignments

She is something of a smart-aleck, and so using the class assignments she decided to create the framework for building this online class. There was no external motivation for this, as the other students in the class were all novices in all aspects of the field of instructional design, technology, and psychology. She was internally motivated by a desire to challenge herself, and to see if she could make the professor laugh at her audacity.

The main assignment was broken into three parts: a learning philosophy, a learning theory, and an instructional theory (Tennyson, Wu, & Hsia, T.-L., in press). This was difficult for her, as the learning philosophy she used in the business world is almost exclusively behaviorist-based. This, she has found, does not match her internal beliefs about how adults learn. She falls firmly in the constructivist/cognitive arena. She had planned to move freely back and forth between the business and academic worlds. However, it was time that she started applying the psychological theorists to her beliefs, and identified what meshed and what conflicted. Thus, she wrote a learning philosophy based in cognitive and constructivist psychology.

Scope

It is important to note here that this main assignment was used to build parts of the online class eventually. In particular, the scope of the project was particularly laid out. Since the class was going to move from an instructor-led, classroom-based class to an instructor-guided/student-guided, online class, decisions needed to be made about what content needed to be transferred to an online book vs. using a textbook. Further decisions needed to be made on how to segment the class into lessons or modules, and what the difference ought to be between these. The scope of the problem included all activities around moving content from an instructor-led format to an online format. This included design decisions like whether to use groups, whether to make the class self-directed or instructor-directed, what content must change in order to be delivered online, and how assessments would change. The scope did not include establishing procedures to be followed in moving courses from instructor-led format to online format. It also did not include developing new content or reviewing textbooks and reading materials for pertinence to the class. There were ample reading materials already developed for the face-to-face version of the class; these could simply be adapted for use online.

Fortunately, this would be a fairly easy class to build. It would be a combination of adapting existing materials and developing new materials to fill in the gap that existed when the lecture piece of the course was removed. The following solution was proposed: develop an online course that, to a great extent, mirrors the classroom-based course. This would be a hands-on course in which participants would (a) read lecture notes, (b) follow along with worked examples, (c) work on practice examples, and (d) work on some problem-oriented situations. The final instructional activity would be designed to help the participant construct a plan for building a real online class, perhaps related to his or her job.

Designing and Building

Initially, there was a gap between the time that the plans for the class were constructed and the okay to move forward was given. The initial plan for the project was determined to be significantly off-base with respect to the calendar. While the lengths of time proposed were accurate for an experienced designer/developer, the actual calendar dates missed the mark by almost six months. Nevertheless, the first design meeting confirmed the above proposed solution, and the designer/developer moved forward.

Syllabus

The first element that was developed was a syllabus. The original syllabus for the face-to-face class was missing a number of elements that the instructional designer deemed necessary for student understanding. These included things like a detailed schedule of readings and assignments, as well as class policies for late work and participation. These elements were added and submitted to the instructor-of-record, who also served as the subject matter expert for the project.

Readings

The second element that was developed was a trial of two different ways of presenting the readings. Moodle contains two different tools for working with tools: books and lessons. The book format appears to be an online book. It is up to the designer to determine what content appears on each page, hopefully paying close attention to the amount so that readers do not have to scroll whenever possible. The lesson format is similar to the book, but allows the instructional designer to add a question at the end of each section, including linking to additional pages for review if the question is not answered correctly. After some discussion with the subject matter expert, it was decided that the class would use the lesson format. This would allow for students to check their reading comprehension and verify higher-level understanding of the material. It would also allow the students to personalize their experience with the class: they would be able to skip around in sections of the lesson as needed, and cover again the material that was more difficult to understand (Wu, Hsia, & Tennyson, in press).

Sequence of Content and Activities

The third element that was developed was the sequence of content and activities week by week. It was determined that the class could be extremely flexible based on the students who were participating. Early in the class, students would complete a self-assessment of their previous knowledge of, experience with, and attitudes towards the content. Based on this assessment, students could be divided into groups according to the instructor's desire in one of three ways. Students were assessing themselves as novices, apprentices, or experts, and so could be divided into those groups. Alternately, if there were enough experts, each expert could work with a group of either novices or apprentices. The third way of dividing students would be to give each expert a group of a few novices and one apprentice. This would allow the experts the opportunity to expand their knowledge by helping less-expert students' work their way through the content. Therefore, activities for each week were developed that could incorporate each way of grouping students. The instructor or the designer could make these activities available based on whichever way the instructor decided to group students.

The Development Process

At this point, the actual development began. Unfortunately, Moodle has a known conflict with Microsoft Word, so simply copying the existing content from Word into the Moodle site was impossible. Every single piece of existing content that was intended to be housed in any non-pdf-format had to be formatted as simple text, entered into Moodle, and then re-formatted to add headings and further guidance for the student. It was unfortunate that this discovery was made by the instructional designer after the schedule for the project had already been set, because this added approximately one month of work, just to make the content usable by students.

Approximately two-thirds of the way through developing the lessons, the instructional designer realized that just including questions and the normal assignments from the face-to-face class would not be enough for students to truly learn the content of the class, such that they would be able to use it on the job. It was determined at this point that a separate group project would be needed, that would add in some practical application techniques (Molkenthin, Breuer, & Tennyson, 2009). The instructional designer decided to add a group project based on the use of a Moodle development site to the class. Specifically, with the group members, each student would explore a Moodle demo site and make use of the tools. They would have access to this Moodle site as though they were an instructional designer (Moodle calls this Instructor Access). The intention of this assignment was to give them practical experience in a safe setting with being an instructional designer/developer and using some of the technology that is available to them. All activities were to be done with the group, and there would be no individual posts. Students were also instructed that the same person should not always post; they should share the responsibilities. Each group assignment associated with this group project directed the students to explore a different part of the Moodle website and determine whether it could or should be used with a project that the group had jointly decided to use. This personalization allowed students to apply the theoretical knowledge they were learning in the rest of the online class, and tie it directly to that knowledge (Schott & Hillebrandt, 2005).

Once the group project was completely built, work on the weekly activities for the online class continued. Perhaps the most difficult part of the development process was creating questions for the lessons (readings) that allowed students to develop higher-level thinking skills regarding the content and allowed students to learn from mistakes made in answering the questions. Additionally, some sections of the lessons did not lend themselves to questions that could be automated. For example, in one of the very first lessons, the question that was developed was "In order to utilize Instructional Design practices, you first need to have a learning problem, and then you need to understand ID methods in order to be able to design a solution." This was developed as a true/false question, based on a section within the lesson that did not use the exact terminology included in the question. The intent was to force the students to read closely, to think about what they were reading, and to begin to make mental connections to practical experience. The original question that was developed required a student to respond in written format, but there was no good, practical way to program Moodle to recognize the variety of possible answers and allow the student to proceed further through the lesson. The best way that we found Moodle to allow written responses is to then have the instructor verify the responses. Unfortunately, this is not practical within the lesson format. Thus, many questions were developed that had multiple choice, true/false, or ranking answers, but were developed in a way that forced the students to read the lesson content closely and then begin making mental connections to practical applications. It is important to note, too, that in most instances, incorrect student responses were met with feedback that both provided the correct response, encouraged students to review again, and allowed them to make personalization choices by either answering the question again or skipping to the next lesson (Spector, Ohrazda, & Van Schaak, 2005).

After building the first two lessons, including questions and accompanying assignments, the instructional designer started to reflect on whether the work she was doing was in line with her basic theories about how students learn. As it happens, she is a cognitivist with a limited belief in constructivism and social constructivism. As it also happens, in her many years of designing and developing training in the business world, she had been forced into a behaviorist model. The authors are certain that most readers are familiar with the business desire to determine a return on investment, especially for training. This means that in developing training in the business world, one strips out everything that doesn't move the goal forward of changing employee behavior, usually on the level of "when they started in class, they couldn't use the tool, and upon finishing, they could use it at a basic, adequate level for their job." Being very used to this model, the instructional designer needed to think carefully about how she was designing the class, and whether she was slipping at all into the too-easy, trained, behaviorist ways that she had used for so many years. It was in her struggle to develop questions that were not simply rote memorization-check, whether questions that she determined that she was not developing a behaviorist-based online class.

At this point, time was progressing, and the due date for the class was approaching almost faster than the instructional designer could build it. She had, she thought, spent too much time struggling with reconciling her theories to the practical build of the online class, and so she compensated by hurrying through the build of the remaining lessons, leaving placeholders for the questions that still needed to be developed. She was able to complete what she considered the shell of the class in a week. She also tends towards perfectionism, so what she called a shell; anyone else would call a complete class. She spent two days verifying that all the elements were in place and that the look and feel of the activities, assignments, and lessons was consistent throughout the course. She then spent one day writing and uploading the missing questions. She had left the addition of the development site on Moodle until later in the term, so that she would know how many students would need to be added. She also left the final sign-off of all the course materials for the final meeting with the subject matter expert, who would also be the instructor for the course.

Implementation

The final meeting was not nearly long enough for all the information the instructional designers wanted to cover. She wanted to walk the instructor through every piece of the course, pointing out all the places that needed instructor input and decisions. Instead, because of time constraints, she restricted her "walk-through" to pointing out the grouping decision materials that the instructor would have to manage, based on the students' self-assessment, and verifying that all else in the course was constructed in a way that the instructor could easily manage the class. Then the semester began, and the students started working. She monitored the class in order to be available to fix any design or development errors that the students or instructor caught, as well as to remind the instructor of the timeframe in which the grouping decision needed to be made. She continued to monitor the class to make sure that students were participating appropriately. Overall, the class was deemed moderately successful, with high student participation and excellent reflection of student learning within the assignments.

CONCLUSION

Designing on-line courses is a complex activity with much of the work involving the transferring of best practices of face-to-face instruction to the on-line course. However, there is a transmediation effect occurring in which some certain features can be transferred successfully and others cannot. The purpose of this chapter was to provide a narrative recollection of this knowledge engineering design problem of adjusting a face-to-face instructional environment into an online learning environment. The computer technology world offers potential practices of instruction that are not available in the face-to-face world of instruction. The goal of the electronic instruction is to employee the best practices of conventional education while employing the unique features of online technology. Much of what can be done online is not research based, but rather intuitive. Our goal from this point on, is to use the developed course as a means for conducting research and evaluating *best practices* for online course design and development. A proposal for this type of research is offered by Tennyson and Jorcazk (2008).

REFERENCES

Lytras, M., Tennyson, R. D., & Ordonez de Pablos, P. (2009). *Knowledge networks: The social software perspective*. Hershey, PA: IGI Global.

Molkenthin, R., Breuer, K., & Tennyson, R. D. (2009). Real time diagnostics of problem solving behavior for business simulations. In Baker, E., Dickieson, J., Wulfeck, W., & O'Neil, H. F. (Eds.), *Assessment of problem solving using simulations* (pp. 205–228). Mahwah, NJ: Erlbaum.

Schott, F., & Hillebrandt, D. (Eds.). (2005). *Outside behavior–inside cognition?* Berlin, Germany: Springer.

Spector, J. M., Ohrazda, C., & Van Schaak, A. (Eds.). (2005). *Innovations in instructional technology: Essays in honor of M. David Merrill*. Mahwah, NJ: Erlbaum.

Tennyson, R. D., & Jorczak, R. L. (2008). A conceptual framework for the empirical study of games. In O'Neil, H., & Perez, R. (Eds.), *Computer games and team and individual learning* (pp. 3–20). Mahwah, NJ: Erlbaum.

Tennyson, R. D., Wu, J.-H., & Hsia, T.-L. (in press). Engaging learning models with information and communication technologies in advancing electronic learning. In Ordóñez de Pablos, P. (Ed.), *Electronic globalized business and sustainable development through IT management: Strategies and perspectives*. Hershey, PA: IGI Global.

Wu, J.-H., Hsia, T.-L., & Tennyson, R. D. (in press). Design strategies for improved online instructional systems. In Ordóñez de Pablos, P. (Ed.), *Electronic globalized business and sustainable development through IT management: Strategies and perspectives*. Hershey, PA: IGI Global.

Chapter 2
Standardization of Virtual Objects

Jordán Pascual Espada
Universidad de Oviedo, Spain

Oscar Sanjuán Martínez
Universidad de Oviedo, Spain

B. Cristina Pelayo García-Bustelo
Universidad de Oviedo, Spain

Juan Manuel Cueva Lovelle
Universidad de Oviedo, Spain

Patricia Ordoñez de Pablos
Universidad de Oviedo, Spain

ABSTRACT

This chapter proposes architecture to unify the development and use of virtual objects. As technology advances more and more "objects" began to appear in digital format, examples include: books, event tickets, airline tickets, agendas, etc electronic purses. These digital objects do not follow a standard format or recommendations since there are no mechanism that allows for treating them in a general way, storing and sharing or being processed by other applications that do not know their format. Based on the problems identified in this document, a proposal is detailed in search for a single structure and for the construction of any virtual object.

INTRODUCTION

The main idea in the Internet of Things is that any "thing" or object, conveniently tagged, may be able to communicate with other objects equally tagged through internet or any other protocols. These objects which are part of the net may contain small chips or embedded systems, depending on their purpose (Kranz, 2010). They may range from home equipment to industrial items or even electrodomestic, cars or even supermarket food. Anything can be tagged to be part of the Internet of things (Kortuem, 2010; Lu, 2008).

DOI: 10.4018/978-1-61520-921-7.ch002

Possibilities of the Internet of things to make people's life easier and to automatize many of our current tasks are huge, for instance, it is possible that the fridge may send an email to our mobile phone if it runs out of milk, we can monitor hospitalized patients by internet... there are lots of practical applications and all of them are seen with a common basis: "things" are communicating with "things" or persons. (Global 2008)

Parallel to the development of technology, more and more objects called "things" which are merely physical start to be seen also in digital format. Examples of them can be seen in: books, maps, e-tickets for gigs, plane tickets, agendas, contact cards, agendas, electronic purses etc.

When we observe the behavior of these digital objects we see there is no standard format or any recommendation to normalize their usage. There is no mechanism by which we can treat them in a general way, store them, share them or process them with other applications which may not know their format.

Problems coming out of this lack of standard format are the following:

- **Difficulties for Decode:** devices with no specific applications to decode the virtual object will not be able to process it. Let's take as an example my Mobile phone; if I transfer a contact card to another user, the Mobile intended to receive it won't be able to decode the incoming information. This handicap leads to the need of installing many applications in case we want to operate with different virtual objects. It makes it harder for a company or developer to place in the market their own virtual object, since nobody would be able to decode it without the suitable software.
- **Lack of Communications:** Ideally, the objects linked to the Internet of things to interact among themselves and with other applications to automate tasks and increase efficiency (Spiess, 2009). Since there is no standard format way to get actions or services from a giving virtual object, it is very difficult to interact with another application. Let`s illustrate it with a cinema ticket which is basically related to being a mere number code with stored information in a company database. The ticket is decoded by a web application and a specific machine. By focusing on this, it is very complicated for a virtual object to directly communicate with other applications or to transfer the ticket to other user.

Internet of things follows the aim of making the Communication between things possible, so things can communicate by themselves with other things and users. A physical thing may have a catalog of actions which is used to communicate, for instance a sensor connected to the net offering service to get position and temperature. The focus of something connected to a web which has a catalog of actions and is able to communicate by itself with other users is crashing frontally against the focusing of virtual objects, which do not exist themselves, independently, as entities, but only to form part of an application which interprets them.

This paper has been divided into the following sections: First, make an analysis of virtual objects in order to define their characteristics and to obtain their commonalities. In a second step we made a proposal for a common structure to support the construction of any type of virtual objects, this structure will be designed around the needs identified. Next we define the possible ways of interaction with standardized objects that have been built according to the proposed structure. At the end of the paper we talk about: potential uses of standardized virtual objects, new lines of future work and conclusions of the investigation.

VIRTUAL OBJECTS

In this document we will give shape to a possible format recommended for the construction of virtual objects. Main objectives are as follows:

- The proposition of a common structure for the construction of virtual object, in which all of them, regardless complexity or Business logic, can be: interpreted equally by any electronic device which is provided with the computational capacity needed (enclosed systems, computers PDAs, mobile phones etc) without the need of any previous configuration or specific software.
- To favour the integration and communications of any virtual object with applications and users. It will be similar to the process followed to integrate physical elements to the internet of things since virtual objects should offer the choice of discovering them to other users or applications, as well as getting their action and service's catalogs or even interacting with them.

To achieve these objectives, the idea is to re-design the virtual objects by using a common structure, turning data interpreted as specific applications into reduced applications that are executed in a safe environment. Devices must know how to read virtual objects and the same should be done to get a cinema ticket, a publicity catalog, a map.This way a single application integrated in the device will be able to decode any virtual object. It also deals with the promoting of objectives related to the internet of things and the communication between objects and between them and applications and users.

Common Aspects for Virtual Objects

It deals with the searching of a unique structure which may lead to the rebuilding, in the same way, very different objects: plane tickets, parking tickets, product catalogs, intelligent publicity, contact cards etc.

In a similar way of a conventional application, parts of a given virtual object could be divided in three layers:

- **Application layer,** in which we include the needed mechanisms so the virtual object can interact with users and applications. The classic form of interactions with users is by means of a rich graphic interface. The interaction with other applications is normally made through a service catalog.
- **Business logic,** in which we found all the Business logic, executable coded or services the object can carry out.
- **Data access layer,** in which necessary data are stored in order to operate with the virtual object.

The design of the structure could be similar to that of a conventional application, but still underlies a great difference in their natures. Frequently, virtual objects are downloaded through internet or are transferred from computers and this is the reason why they should offer a structure very capable of migrating between devices in a very dynamic way, very lightly, and with no installation required. Instead of settling should be run in a "sandbox" with limited permissions (Yeon-Seok, 2007).

Generally, all objects should have a series of common properties which may allow identification: name, type and identification.

At the time of dealing with an object, there are common actions which are similar to those of a file. Can they be copied? Can they be modified? Can they change owner, or be transferred?

All needs and observations commented, have been taking into count when it comes to elaborate a common structure for virtual objects.

PROPOSAL: STRUCTURE OF VIRTUAL OBJECTS

Being based upon detected needs, this proposal defines that the internal structure of a virtual object is formed by a group of files of different nature. The types of files which form the structure are the following:

- **Descriptor:** XML file, which contains information about identity, configuration, general behaviours, arrangement and execution of the virtual object.
- **Graphical interfaces:** XML files, each one represents a screen, which is the means by which the user can communicate with the logic of a virtual object.
- **Service catalogs:** there are resources which have the function of showing the applications or programmes, the actions an object may carry out. This interaction is achieved by means of service catalogs, which execute actions in the business logic of the object.
- **Executable code:** a file which contains the needed code to execute the virtual object logic. The code can be obtained in different formats or languages of programming in order to be executed in devices of different characteristics.
- **Data storage:** if the logic regarding Business services may require persistence, this must be provided with a file responsible for the arrangement of that information. Virtual objects are relatively simple models when compared to that of a conventional application. They are not thought to store a great deal of registers but only a few values.
- **Additional resources:** Could be included a Lumber of extra resources, in a non-limited way. Generally, this will be multimedia resources: images, icons and videos, which will be used to complete the graphical parcel regarding virtual object.

Development of Virtual Objects

To illustrate the use of virtual objects has implemented a movie ticket, following the proposed specification; this prototype will be used to illustrate the structure and operation of virtual objects.

Complementing the virtual object input has developed an application "manager of virtual objects," which runs on the Android mobile operating system (Android, 2010).

The ideal container for virtual objects is an electronic device that we carry with us all day, which allows us to interact with objects at anywhere. We selected the Android mobile platform for developing the prototype because it is open source and devices that have this operating system have characteristics of computing and communication technologies, these features are sufficient to ensure interaction with virtual objects. New technical developments make it increasingly small mobile devices converge to computers so that in a few years, the interaction with virtual objects would be possible from all phones (Chang, 2009).

This application is able to interpret and work with any virtual object built to specification.

Business Logic: As a first step to begin the development of the entry, we will analyze the business logic, the actions that the object could carry out.

- Display information about the movie: title, synopsis, images and videos.
- Show information about entry: film, room, position in the room schedule.
- Validate against the receiver installed in the door of the cinema, to gain entry. In this case, the virtual object entrance is modeling a real object of considerable value as would be necessary to introduce elements of validation. According to the structure of virtual objects such mechanisms are im-

Code 1. Virtual Object ticket Java Class

```
public class ticket extends VirtualObject{

    private String filmName="RobotMovie";
    private int cinema = 43223;
    private int chair = 34;
    private String[] pictures= {"pic1.png","pic2.png","pic3.png"};
    private int currentPicture = 0;

    public ticket(){
        super();
    }

    public boolean useIt() throws Exception{
        // Data warehouse access. Key - used.
        boolean isUsed = loadDataBoolean("used");if(!isUsed){
            // Connect to the server of the cinema
            BluetoothAdapter mBluetoothAdapter;...

            // Modify data warehouse. Key - used, Value - true
            saveDataBoolean("used",true);return true;} else {return false;}}
    public String getFilName(){return filmName;}
```

plemented within the code associated with this function (Tripathi 2010).

The business logic is implemented in a code file as if it were a conventional application. The virtual object can contain multiple source files that implement the same business logic, so that the devices running the object select the appropriate code to run on the operating system. This prototype uses the Java language for implementation, because we know that only runs on Android system. The implementation is done in standard Java by inheriting the VirtualObject specific class. The code will be dynamically loaded by the manager of virtual objects, so that methods to be invoked must be declared as public.

Data: Sometimes the business logic may require that some data have a persistent nature. Virtual objects are simple models, so they will not store many registers and will not require a traditional database. To support persistent data they are declared as key-value pairs in a specific file. From the executable code one can easy access to the stored values, using special methods, which are implemented in the class VirtualObject. This system achieves a simple and efficient synchronization with the data store that is almost transparent to the programmer. In the case of the cinema ticket that we are implementing the value "used" would be a persistent value.

Interfaces: at this point we must analyze ways of communication that the virtual object will have with other users or applications. The proposed virtual objects, offers three different interfaces:

- **Graphic interfaces:** is the main form of interaction with the virtual object, its use is to provide a simple visual environment to enable communication between user and virtual object. Through the interaction on screen the user can interact with the object and take action, to conduct executions in the associated code. The user interfaces should belong to one of two types:

Figure 1. Communication mechanisms of virtual objects

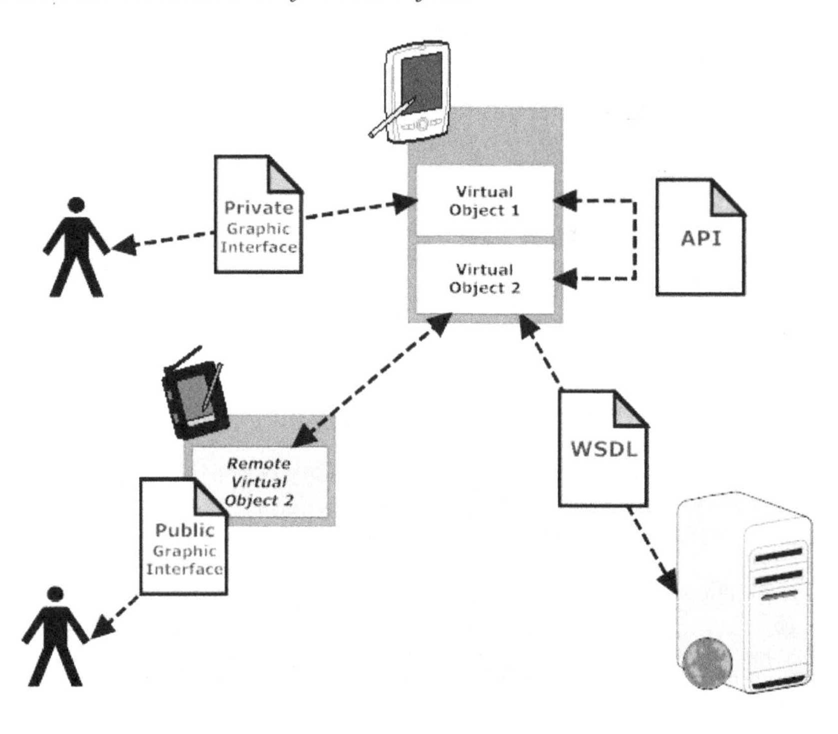

    ◦ **Private:** used by the user who owns the virtual object from your device.
    ◦ **Public:** can be added if you want other users to discover and connect to the virtual object remotely.
- **WSDL:** stands for Web Services Description Language. Its mission is to show the public services the object you want to publish so that other applications can access them via the Internet.
- **APIs:** is a service descriptor with a similar purpose to the WSDL, the difference is that public services described in it can only be consumed by other virtual objects that are housed in the device.

Graphic Interfaces: Each XML file refers exclusively to a screen that can be displayed during performance of the virtual object. It is possible that in the course of implementation several transitions may occur between screens, so there may be several files of frontend.

To describe the elements that appear on the screens and how they behave, we have started from a smaller version of the syntax used by the Android system (User Interface, 2010), describing user interfaces in a relative way, so they can be interpreted in the same way regardless of the resolution or screen size of the device.

From the source code we will make transitions between screens, or modify the properties and behavior of its elements.

In the case of the cinema ticket it has included a private graphical interface, through which the owner of the ticket could manage, and publish a graphical interface that allows other users to connect to the virtual object entry and see an overview of film.

Configuration: The configuration file is an XML document which contains information about the identity, configuration and implementation of the virtual object. The information contained is as follows:

Figure 2. Graphic Interface file

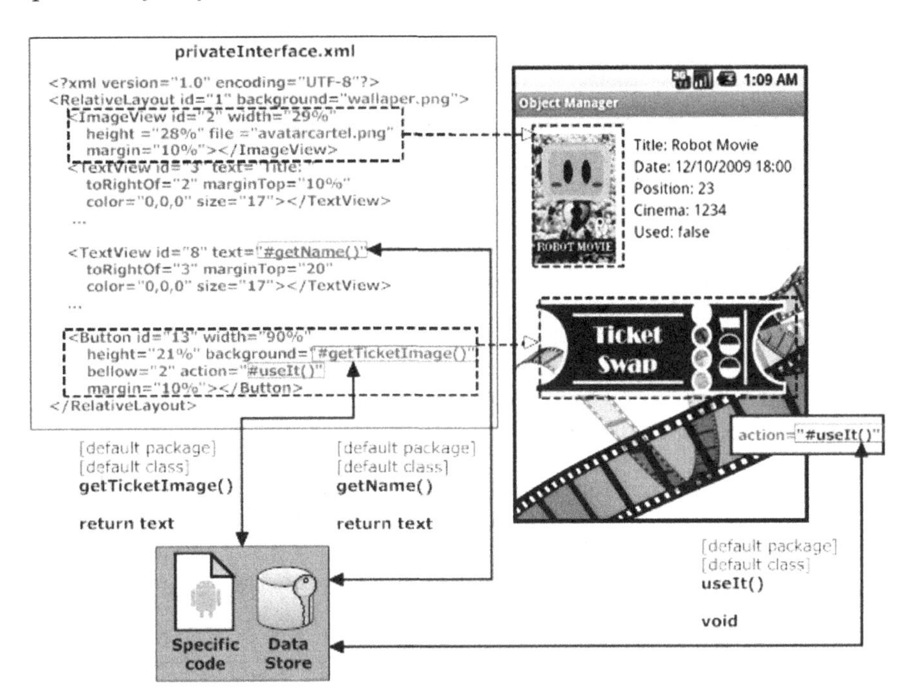

Code 2. Java code for changing graphic interface public

```
public void nextScreen(){
    setPublicContentView("publicInterface2.xml");}
```

- **Identity Object:** Name, Type, Unique Identifier (if it is unique) and icon.
- **Behaviour of the object:** If it is transferable, copyable or editable, if you have an expiry date…
- **Interfaces:** Name the main interfaces of the object.
- **Executable code:** Name of the source files that may exist, at least there must be one.
- **Data:** Name of file data store, and class, if optional. In the particular case of this movie ticket could be a valid configuration file:

Every single virtual object cannot have two equal entries, so that's why it is assigned a unique type of identifier. For the same reason it can't be possible to copy or edit the object, but instead we can decide the virtual object entry itself that could be transferred to another device if we want, so the positive attribute is transferable. Interfaces which hold a virtual object can be declared, depending on the function of the object possess one or the other, in case of graphics interfaces we will only declare which is the main screen, which must be loaded when we initiate the object, either in a local graphic interface or even in a remote public graphic interface.

The element "code" serves to declare the packages containing source, may contain several, and so that the virtual object manager select the right one for your operating system. Optionally you can declare which is the main class in the code. There are also disclose the name of the data

Code 3. Configuration file of the virtual object.

```
<?xml version="1.0" encoding="UTF-8"?>
<virtualObject>
    <name>Ticket: Robot Movie</name>
    <type>Ticket</type>
    <id>47236271</id>
    <transferable>YES</transferable>
    <expiration>12-10-2009</expiration>
    <editable>NO</editable>
    <copy>NO</copy>
    <interface
        private="privateInterface.xml"
        public="publicInterface.xml"
        wsdl="interface.wsdl"></interface>
    <code
        android="robotMovie9780.apk"
        j2me="robotMovie9780.jar"></code>
    <data>info.obj</data>
    <mainClass>com.robotmovie.Ticket</mainClass>
    <icon>icon_tiket.png</icon>
</virtualObject>
```

warehouse, if you have it will not be necessary in all virtual objects.

Resources: Additional resources will be any type of files which are needed to support the use of virtual object, it usually consist of images or other types of multimedia material used in graphical interfaces.

USING VIRTUAL OBJECTS

Virtual objects are interpreted by an application which we call Virtual Object Manager, which is responsible for managing the virtual objects within a device. It must be designed to be installed and run on the given device, taking into account their characteristics, operating system, or programming languages it supports. Objectives of the manager are:

- **Load, interpret and run** any virtual object that has been built following a proposed structure. The first step to start using an object is to pick a manager, selecting the object configuration file. Once loaded the user can start interacting with the object, the manager interpret their interfaces and execute the corresponding code.
- **Store and manage** virtual objects. We will often use multiple virtual objects simultaneously; the manager must provide mechanisms to store and manage. Virtual objects that were loaded on the device must be displayed in an orderly manner to the user, so that it can interact with them and manage them, that is: delete, copy or transfer provided that the nature of the subject permits.
- **Publish** virtual objects and service catalogs. It can be specified in the logic of the object that allows execution of remote or publish their services, the manager has mechanisms to support such protocols relying on Bluetooth (Bluetooth, 2009), or other protocols such as: Internet, wireless etc.

Figure 3. Screenshots: Interpreting charged object

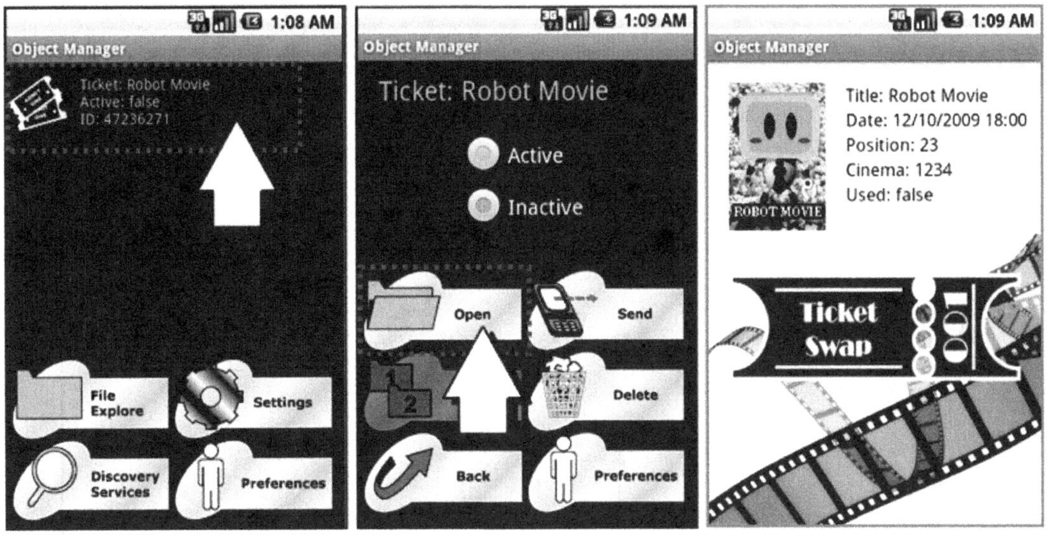

- **Discovery and remote execution.** The manager will be able to discover the virtual objects that other devices publish. Once discovered the virtual objects may be performed remotely

Local Execution

Once the object is loaded into the device, the user can select and interpret it, the result is a rich graphical interface through which users conduct invocations to the services included in the executable code of the object (figure 3).

1. When you open a virtual object it is loaded into main screen graphic interface. If the user manipulates one of the interface elements, such as a button, it can launch events that require the execution of the code. In the case of entry, within the private interface Virtual Object Main Entrance, pressing the "Ticket Swap", launches an event that requires execution of code.
2. When we launch an event, the manager compiles the corresponding information to the executable action, name of the method and parameters and code files where it is contained.
3. The manager will access the database and the code that has been formerly indicated, will search and will execute the method. In case that the method has a way to return, it will be stored. In the example of the cinema ticket, the executable code will synchronize through Bluetooth, with the cinema embedded system, will interchange security codes and if the ticket hasn't been formerly validated, it will tag it as used and will gain us entry to the cinema (figure 5).
4. The last step for every execution consists of refreshing the screen, the changes on the Graphics interface which will be done from the executable code, so that every time we execute a method a change could have been done.

During the interaction of a user with a virtual object, four stops are repeated in a cyclical way.

Figure 4. Diagram of local execution process

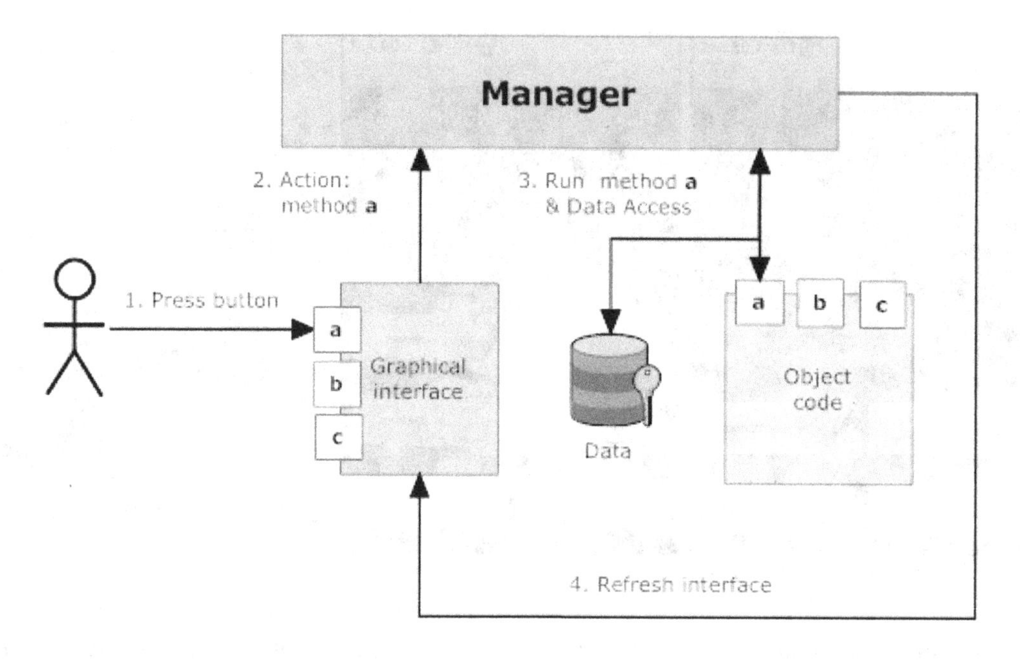

Figure 5. Enforcement action shot from the graphic interface

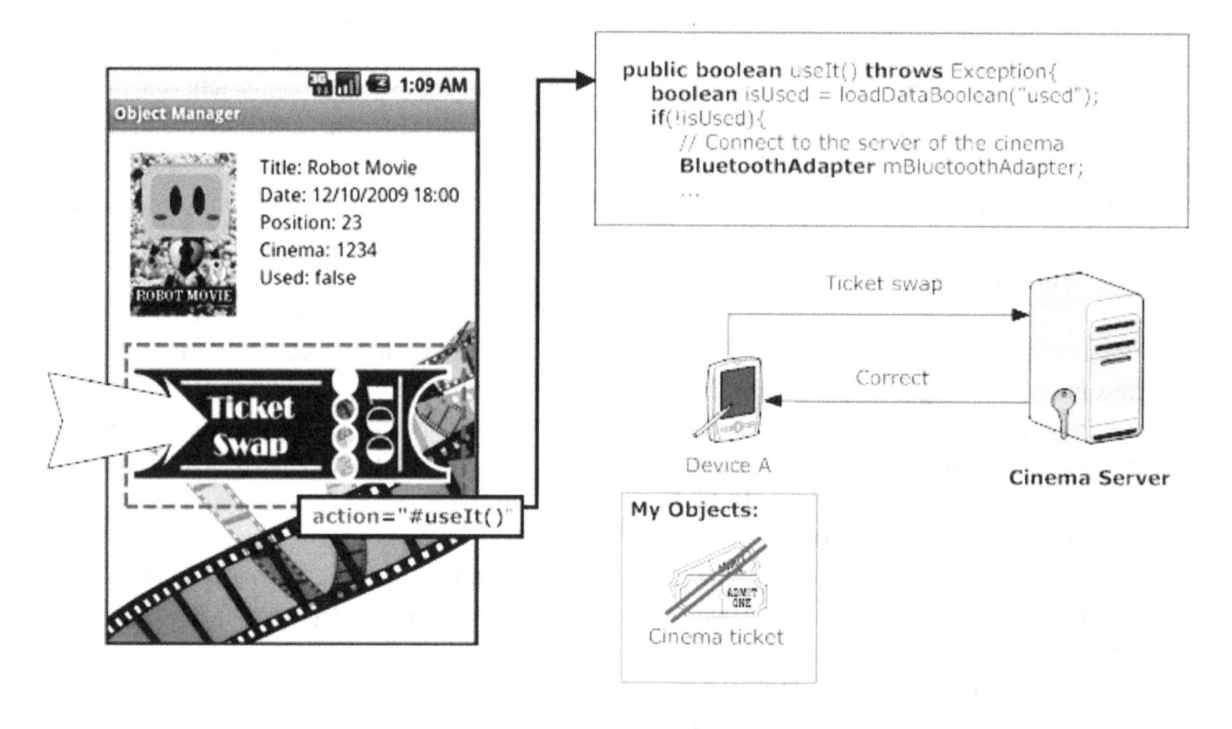

Figure 6. Diagram of remote execution process

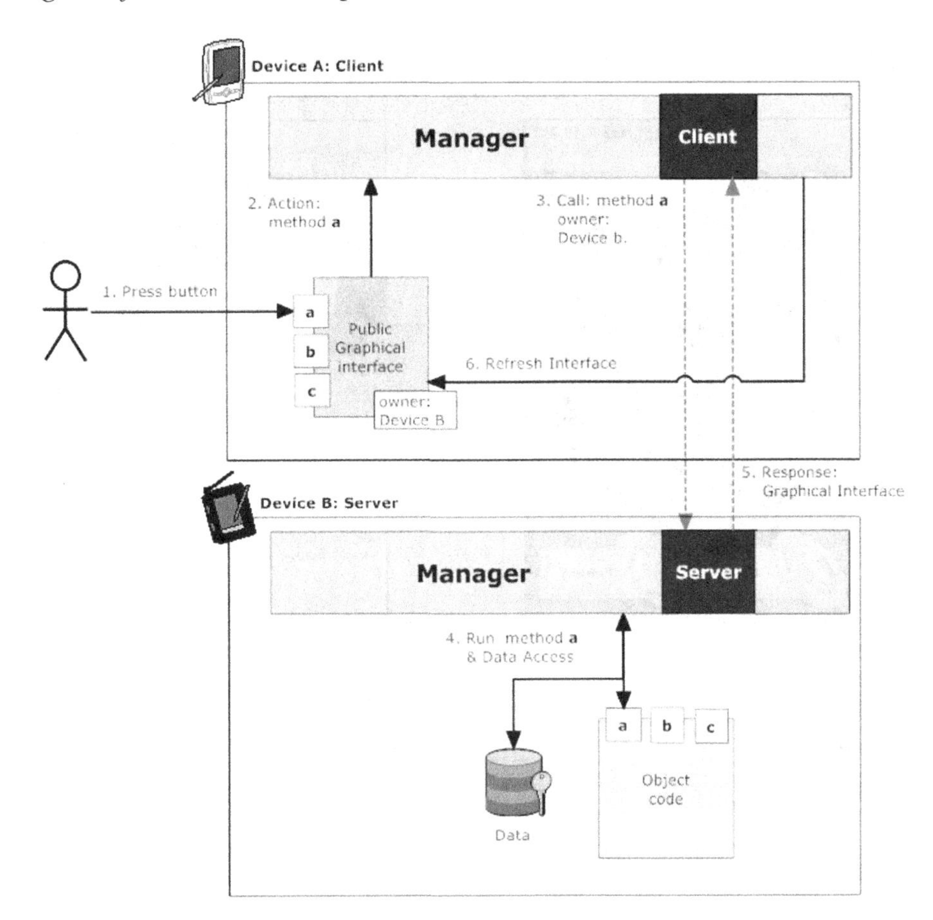

Remote Execution

Depending on the logic of the virtual object is possible that this has to be used by more than one user, to support this form of use remote execution is being included, which enables a device to act as a server of a virtual object, while other users connect to it and execute it remotely and so discovering virtual objects that are loaded and active in other devices.

To start remote execution virtual objects must be discovered by other devices via Bluetooth, Wifi and other protocols. Once discovered the virtual objects could be executed remotely by other client devices, so that the object code will run on the device server and it will exchange a flow of data with the client device the data exchange.

1. When you open one of the virtual objects discovered in client device, the screen shows the main graphic public interface, which has been previously sent, like all resources associated to it. If the user manipulates one of the interface elements, such as a button, it can launch events that require the execution of a code. In the particular case of input to connect a remote client to the object, displaying the main public interface, which shows the synopsis of the movie by clicking on the image above it triggers an event that requires execution.
2. Whenever an event is launched in the graphic interface manager collects information for the action to be executed, similarly to local implementation.

Figure 7. Remote execution process

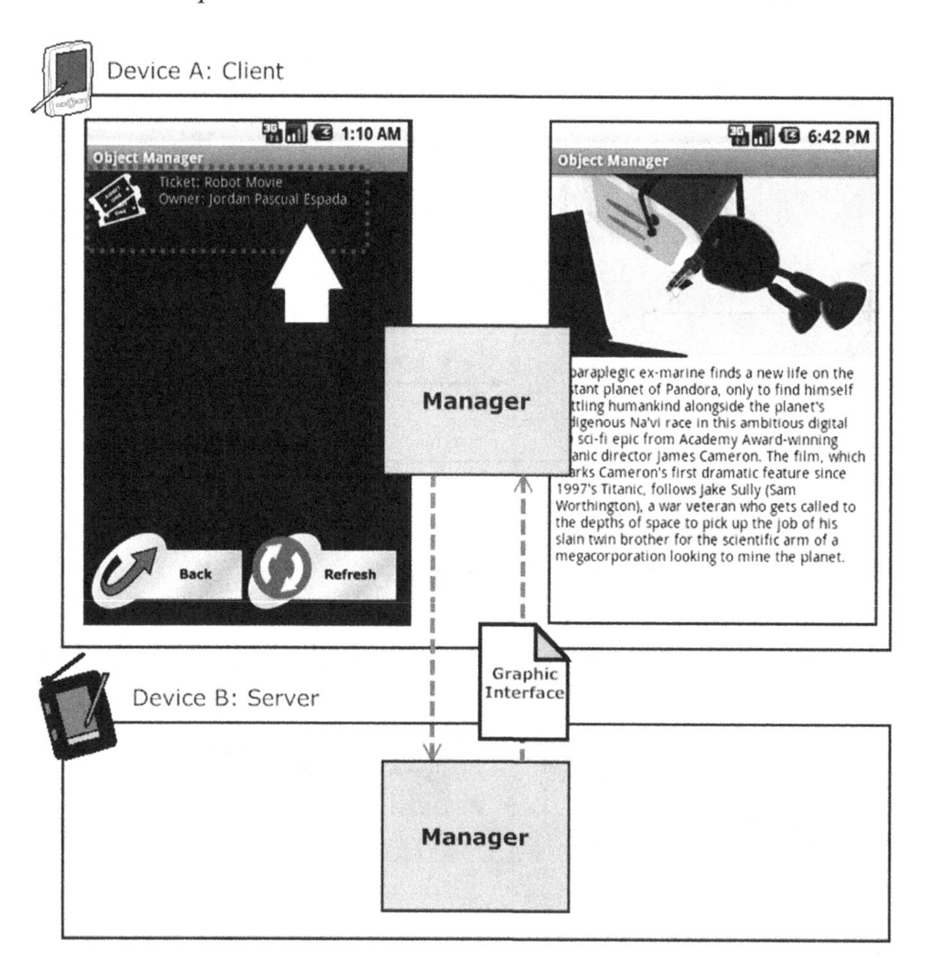

3. The Manager sends a call to the device manager server, indicating what action to take.
4. The device manager server is synchronized with the data store of the virtual object,.It finds and executes the method in the code file that has been indicated in the received request. Similarly, as if it was running a virtual object locally. Changes in public graphical interface are also made from the executable code, so each time you run a method a change could have been possibly made. The manager asks the code through predefined methods which is the current public graphic interface.
5. In the specific case of the input image which is shown in the description it has changed so it is necessary that the client load a new interface.
6. The Server sends an answer to the client which contains the public Graphics interface to be shown, according to the action that has been executed. After getting the answer from the Server, the client should refresh the graphic interface which is shown. In order to do that, the manager loads the public graphic interface which has been received.

During the interaction of a user with a virtual object four last stops are repeated cyclically.

POTENTIAL USES

Nowadays, many objects, either real or virtual, are good candidates to be re-designed as virtual objects. Thanks to these changes we could an improvement in the lifestyle of people, making it easier and automatizing many daily life tasks. Some examples are the following:

- **Tickets:** cinema tickets, theatre, events, plane tickets, underground or train, parking lot tickets. The proposal for virtual objects offers the possibility of storing them, for instance in our Mobile or PDA, as if it was our digital Purse.
- **Multimedia objects:** product catalogs, intelligent publicity, contact cards. The structure proposed offers the possibility of creating Rich interfaces, including lots of control, in sound, videos etc.
- **Application generated resources:** Shopping lists, events and schedules, since it contains business logic and services, a list cannot be limited to a series of numbered products, it can be also provided with certain intelligence and capacity to interact with other applications or elements.
- **Collaborative objects, blogs and others.** The virtual objects may include the possibility of being accessed in a remote way, that is why a service may interact with different clients.
- **Remote control:** the remote execution and the ability of building up Rich graphical interfaces gives us the Choice of connecting from a device to another ones which are not really virtual objects but that have programmed public interfaces and services. This way a single device could act as a remote control for a great number of elements.

FUTURE RESEARCH DIRECTIONS

Future work is oriented towards the optimization and Benefit of the target structure and defined behaviour for the virtual object. It is also directed to the study of new ways of integration and synchronization with other applications and objects, either physical or virtual. Future investigation is divided in different areas:

- To optimize and debate the target structure for the virtual objects, with the aim of making them simpler and more efficient to be built, freeing the Developer off a greater number of tasks.
- Investigation, perfection of new ways of Communications and synchronization of virtual objects with external applications and other objects, so as to optimize the integration. The adaptation of virtual objects to the dominant technology in the world of mobile communications in order to increase efficiency and capacity of interaction (Thomas, 2010).
- Security can be a key point in certain types of virtual objects. All information and business logic of the object is in files to which the user can access. What makes a multiplicity of zones in which the object could be modified or corrupted.

Much of the security on objects that required it would fall on the developers of the Object, for example, in especially important virtual objects would be advisable to duplicate the information on an external server or "in the Cloud" and perform validation against it. It implies greater complexity in the development of object code. A new trend of investigation consisting of integrating normalized virtual objects with simple Cloud computing techniques (Sedayao, 2009).

- The logic of Business in virtual objects is coded in different languages depending on the manager device that is why it is useful to include several codifications of the same logic in order that the object code can be interpreted in devices of different operative Systems. Future research will be directed to treat the object implemented in a specific language can be executed within the manager of objects of any type of device.

- To study new potential usage and fields for the application of virtual objects as well as the impact on users and Business.

CONCLUSION

The benefits of the proposal on the current solution in the modeling of virtual objects are:

- It unifies the way we build virtual objects with a concrete structure, Strong enough to model complex virtual objects.
- It is designed so devices with the suitable computation ability can be executed or correctly interpreted in any normalized virtual objet, it doesn't matter the object logic or the device characteristics, (operative system, resolution...).
- Easy development, using languages of general purpose and widely extended formats for the construction of virtual objects. It offers automatic support to main properties which may define the object behavior.
- Strengthens and makes it easier the Communications between virtual objects, users and application, the same with the transfer or interchange of virtual objects.

REFERENCES

Android. (2010). *Specifications*. Retrieved from http://developer.android.com/

Bluetooth Core Specification v3.0 + HS (2009). *Specifications*. Retrieved from http://www.bluetooth.com/Bluetooth/Technology/Building/Specifications/

Chang, Y., Chen, C. S., & Zhou, H. (2009). Smart phone for mobile commerce. *Computer Standards & Interfaces*, *31*(4), 740–747. doi:10.1016/j.csi.2008.09.016

Global Trends 2025: A transformed world. (2008). *Appendix F: The Internet of things* (background). SRI Consulting Business Intelligence.

Kortuem, G., Kawsar, F., Sundramoorthy, V., & Fitton, D. (2010). Smart objects as building blocks for the Internet of things. *IEEE Internet Computing*, *14*(1), 44–51. doi:10.1109/MIC.2009.143

Kranz, M., Holleis, P., & Schmidt, A. (2010). Embedded interaction: Interacting with the Internet of things. *IEEE Internet Computing*, *14*(2), 46–53. doi:10.1109/MIC.2009.141

Lu, Y., Yan, Z., Laurence, T. Y., & Huansheng, N. (Eds.). (2008). *The Internet of things: From RFID to the next-generation pervasive networked systems*. Auerbach Publications, Taylor & Francis Group.

Sedayao, J., Su, S., Ma, X., Jiang, M., & Miao, K. (2009). *A simple technique for securing data at rest stored in a computing cloud*. (LNCS 5931), (pp. 553-558).

Spiess, P., Karnouskos, S., Guinard, D., Savio, D., Baecker, O., Souza, L. M., & Trifa, V. (2009). *SOA-based integration of the Internet of things in enterprise services*. IEEE International Conference on Web Services, ICWS 2009, (pp. 968-975).

Thomas, J. P., Wiechert, F., Florian, M., Patrick, S., & Elgar, F. (2010). *Connecting mobile phones to the Internet of things: A discussion of compatibility issues between EPC and NFC*. Auto-ID Labs.

Tripathi, A., Suman Kumar Reddy, T., Madria, S., & Mohanty, H. (2009). Algorithms for validating e-tickets in mobile computing environment. *Information Sciences*, *179*(11), 1678–1693. doi:10.1016/j.ins.2009.01.018

User Interface Android. (2010). *Specifications*. Retrieved from http://developer.android.com/guide/topics/ui/index.html

Yeon-Seok, K., & Kyong-Ho, L. (2007). A lightweight framework for hosting Web services on mobile devices. *Proceedings of the 5th IEEE European Conference on Web Services*, ECOWS 07, (pp. 255-263).

Chapter 3
Forum Summarization to Support Tutor and Teacher in Group Interaction Management

Antonella Carbonaro
University of Bologna, Italy

ABSTRACT

The process of summarizing information is becoming increasingly important in the light of recent advances in resource creation and distribution and the resulting influx of large numbers of information in everyday life. These advances are also challenging educational institutions to adopt the opportunities of distributed knowledge sharing and communication. The chapter presents a summarization system to support tutor in managing student communication and interaction within a learning framework. Results show the adequacy of the system in identifying a good content summarization and then in improving the efficiency and effectiveness of the context in which summarization can be integrated.

INTRODUCTION

The Internet has grown beyond merely hosting and displaying information passively. It provides easy access for people to share, socialize, and interact with one another. Information displayed and exchanged between people are dynamic, in contrast to static information depicted in the older age of the Internet.

DOI: 10.4018/978-1-61520-921-7.ch003

Forums are web virtual spaces where people can ask questions, answer questions and participate in discussions. The availability of vast amounts of thread discussions in forums has promoted increasing interests in knowledge acquisition and summarization for forum threads. Forum thread usually consists of an initiating post and a number of reply posts.

Text summarization has been an interesting and active research area since the 60's. The definition and assumption is that a small portion or several keywords of the original long document can represent the whole informatively and/or indicatively. Reading or processing this shorter version of the document would save time and other resources (Zhou, 2006). This property is especially true and urgently needed at present due to the vast availability of information.

Moreover, the Web is moving toward a social place and increasingly producing new applications: there has been a shift from just existing on the Web to participating on the Web. Community applications and online social networks have recently become very popular, both in personal/social and professional/organizational domains (Kolbitsch, 2006). Most of these collaborative applications provide common features such as content creation and sharing, content-based tools for discussions, user-to-user connections and networks of users sharing common interest, reflecting today's Web 2.0 rich Internet application-development methodologies. Concept-based systems to facilitate knowledge representation and extraction and content integration are obtaining a great deal of interest (Bighini, 2004).

Concept-based approach to represent dynamic and unstructured information can be useful to address issues like trying to determine the key concepts and to summarize the information exchanged within a personalized environment, for example within a technology-enhanced learning system. Indeed, a virtual learning system is not only a set of contents anymore, but also may include a collaboration spaces and tools such as forums, chats or shared document areas. To support automatic analysis of learner's progress in terms of the knowledge structures they have acquired, different methodology can be used. For instance, it could be useful to automatically construct concept maps or domain ontologies based on the messages posted to online discussion forum.

Interaction in Learning Environments

The amount of interaction in technology-enhanced learning systems appears to be an important element of learning effectiveness. Wagner (1994) defined interaction as an interplay and exchange in which individuals and groups influence each other. Thus interaction focuses on the interpersonal behaviors in a learning community. Gunawardena and Zittle (1997) argued that on-line students can create social presence by projecting their identities and building on-line communities through text-based communications alone.

Rovai and Barnum (2003) also provided evidence that students' perceived that learning from on-line courses was positively related to quantitative measures of course interaction. However, judgments about the relative importance of the two interaction variables are difficult because these variables are correlated. Nonetheless, only the active interaction measure, representing for example the number of student message posted to discussion boards or the number of participation in forum thread, was significant. This finding affirms the importance of providing opportunities for on-line students to learn by active interaction with each other and with the instructor (Zirkin, 1995). Consequently, educators should develop and include highly interactive material in distance learning and encourage students to participate in on-line discussions. Findings also suggest that passive interaction, analogous to listening to but not participating in discussions, was not a significant predictor of perceived learning in the present study. Therefore, using strategies that promote active interaction appears to lead to greater perceived learning and may result in higher levels of learner satisfaction with the on-line learning environment. The quality of interactions is another important aspect of communications that should be the topic of further research and goes over the objective of actual work.

So, it is necessary to support tutors in order to manage the communication services provided by the community and to monitor student interactions. This aspect has been largely neglected in the literature. However, supporting tutors is very important to make learning communities effective.

Although some platforms offer reporting tools, when there are a great number of students and a great diversity of interactions, it becomes hard for a tutor to extract useful information. Conceptual-based techniques can build analytic models and uncover useful information from data.

The system we want to propose in the chapter can find application in any context in which the group interaction is a requisite, and we believe that a Web-based learning system is an ideal application domain. To the best of our knowledge, no systems use concept-based approach to represent online forum information in a learning environment summarizing text to represent the whole forum content (as opposed to creating a summary of each message in a thread individually as in (Farell, 2002). In this research we propose an extractive method, that is the summaries that we generate use only text that already exists and their content is determined by the data combined with algorithms and heuristics.

SUMMARIZATION

Summarization is a widely researched problem. As a result, researchers have reported a rich collection of approaches for automatic document summarization to enhance those provided manually by readers or authors as a result of intellectual interpretation. One approach is to provide summary creation based on a natural language generation (as investigated for instance in the DUC and TREC conferences); a different one is based on a sentence selection from the text to be summarized, but the most simple process is to select a reasonable short list of words among the most frequent and/or the most characteristic words from those found in the text to be summarized. So, rather than a coherent text the summary is a simple set of items.

From a technical point of view, the different approaches available in the literature can be considered as follows. The first is a class of approaches that deals with the problem of document classification from a theoretical point of view, making no assumption on the application of these approaches. These include statistical (McKeown, 2001), analytical (Brunn, 2001), information retrieval (Aho, 1997) and information fusion (Barzilay, 1999) approaches. The second class deals with techniques that are focused on specific applications, such as baseball program summaries (Yong Rui, 2000), clinical data visualization (Shalar, 1998) and web browsing on handheld devices (Rahman, 2001). (NIST) reports a comprehensive review.

The approach presented in this paper produce a sets of items, but involves improvements over the simple set of words process in two means. Actually, we go beyond the level of keywords providing conceptual descriptions from concepts identified and extracted from the text. We propose a practical approach for extracting the most relevant keywords from the forum threads to form a summary without assumption on the application domain and to subsequently find out concepts from the keyword extraction based on statistics and synsets extraction. Then semantic similarity analysis is conducted between keywords to produce a set of semantic relevant concepts summarizing actual forum significance.

In order to substitute keywords with univocal concepts we have to build a process called Word Sense Disambiguation (WSD). Given a sentence, a WSD process identifies the syntactical categories of words and interacts with an ontology both to retrieve the exact concept definition and to adopts some techniques for semantic similarity evaluation among words. We use GATE (Cunningham, 2002) to identify the syntactic class of the words and WordNet (Fellbaum, 1998), one of the most used ontology in the Word Sense Disambiguation task.

GATE provides a number of useful and easily customizable components, grouped to form the ANNIE (A Nearly-New Information Extraction) component. These components eliminate the need for users to keep re-implementing frequently needed algorithms and provide a good starting point for new applications. These components implement various tasks from tokenization to semantic tagging and co-reference.

WordNet is an online lexical reference system, in which English nouns, verbs, adjectives and adverbs are organized into synonym sets. Each synset represents one sense, that is one underlying lexical concept. Different relations link the synonym sets, such as IS-A for verbs and nouns, IS-PART-OF for nouns, etc. Verbs and nouns senses are organized in hierarchies forming a "forest" of trees. For each keyword in WordNet, we can have a set of senses and, in the case of nouns and verbs, a generalization path from each sense to the root sense of the hierarchy. WordNet could be used as a useful resource with respect to the semantic tagging process and has so far been used in various applications including Information Retrieval, Word Sense Disambiguation, Text and Document Classification and many others.

Noun synsets are related to each other through hypernymy (generalization), hyponymy (specialization), holonymy (whole of) and meronymy (part of) relations. Of these, (hypernymy, hyponymy) and (meronymy, holonymy) are complementary pairs. The verb and adjective synsets are very sparsely connected with each other. No relation is available between noun and verb synsets. However, 4500 adjective synsets are related to noun synsets with pertainyms (pertaining to) and attra (attributed with) relations.

The subset of keywords related to each thread forum helps to discriminate between concepts. In such a way, two texts characterized using different keywords may result similar considering underling concept and not the exact terms. We use the following feature extraction pre-process. Firstly, we label occurrences of each word as a part of speech (POS) in grammar. This POS tagger discriminates the POS in grammar of each word in a sentence. After labelling all the words, we select those ones labelled as noun and verbs as our candidates. We then use the stemmer to reduce variants of the same root word to a common concept and filter the stop words.

A vocabulary problem exists when a term is present in several concepts; determining the correct concept for an ambiguous word is difficult, as is deciding the concept of a document containing several ambiguous terms. To handle the word sense disambiguation problem we use similarity measures based on WordNet. Budanitsky and Hirst (2001) give an overview of five measures based on both semantic relatedness and semantic distance considerations, and evaluate their performance using a word association task.

We considered two different similarity measures; the first one is proposed by Resnik (1995) while the second is proposed by Leacock-Chodorow (1998).

The Resnik similarity measure is based on the information content of the *least common subsumer* (LCS) of concepts A and B. Information content is a measure of the specificity of a concept, and the LCS of concepts *A* and *B* is the most specific concept that is an ancestor of both *A* and *B*. The Leacock and Chodorow similarity measure is based on path lengths between a pair of concepts. It finds the shortest path between two concepts, and scales that value by the maximum path length found in the is–a hierarchy in which they occur. We propose a combined approach based on the two described measures considering both a weighted factor of the hierarchy height and a sense offsets factor. More detailed information can be found in (Carbonaro, 2010).

System Architecture

The following sentence enables us to illustrate system architecture and its functionalities.

The Semantic Web is an evolving development of the World Wide Web in which the meaning (semantics) of information and services on the web is defined, making it possible for the web to understand and satisfy the requests of people and machines to use the web content

After several transformations and reductions of the input text, the result is the following seman-

Figure 1. The process information flow

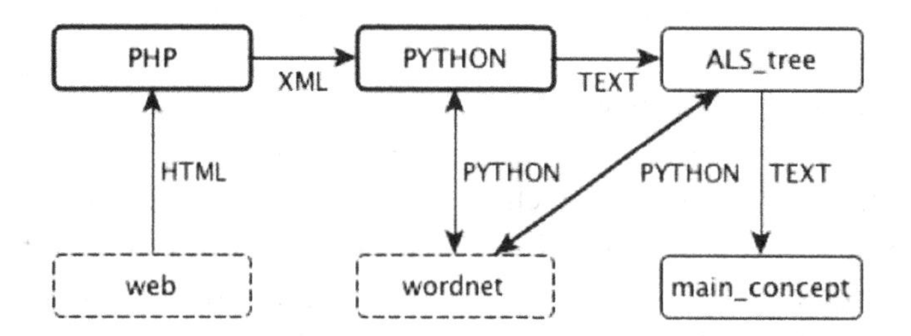

tics map representing the terms and the relative frequency. Depending on their frequency, we can consider the term more or less important within the context. Each term is reduced to common root using a stemming process. {satisfy, 1; web, 5; semantic, 2} is the result obtained using the list of terms and their frequency.

The next step is to reduce the list according to term frequency: we calculate the average value of frequencies and we discard the terms corresponding to the frequency below the average. In fact, we do not want to completely delete the results obtained up until now, but we would like to offer to the end-user different semantic maps relating to the different stages of reduction and refinement of the text.

Last process step is more complex but very effective. For each remaining term we evaluate its synset, that is names, verbs, adjectives and adverbs grouped into sets of cognitive synonyms. Each of these is compared with the others 'parent terms' so verifying the existence of a conceptual link between analyzed words. If this succeeds, we delete an entire branch of the tree, otherwise, the process continues with subsequent comparisons.

This last stage is recursive so to prune more possible branches and to obtain a reduced and significant set of terms. The project is structured so that any user, while not having read the text of the post, understand the concept underlying the message.

In this particular case the computation reduced to the minimum terms the initial tree ({semantics}) trying to extract a single node representing the main concept of the example sentence.

We chose to implement the various text processing using the W3C standard. The content acquisition phase is realized entirely in PHP.

The pre-processing phase produces a XML file based on RSS formats. The main core of the semantic process is a script developed using python programming language. This script processes the textual content and produces a semantic tree. The tree is consecutively and recursively reduced using a set of WordNet libraries. The reduction is organized as follows: the input consists of terms selected on their frequency, for each term we evaluate corresponding synset. The terms of each list are compared between them to find any semantic matches. We delete the branch that contains the actual term and we also discard the parent node, that is the synset generator. The elaboration ends providing for each post meaningful words that summed up the meaning. The difference between a tag cloud or a list of words sorted according to the frequency is in the semantic reduction of tree that uses the meaning of the words to conceptually sort them inside the tree.

Figure 1 shows the whole process information flow. The first computation is a php-based reading of the web content. Next, we transfer XML document to the python compiler that, through the use of WordNet and Natural Language Toolkit libraries, returns the reduced tree in a text-based format (ALS_tree, Lexical and Semantic Analisys tree). Successively, we extract, using WordNet and python language, the main concept representing analyzed text. The main concept contains one or more meaningful phrases.

We show the several procedures and the order we used to produce the key terms summarizing main concepts of the analyzed forum. The first one is devoted to stop-word elimination, to eliminate generic words with no semantics and which do not aggregate relevant information to the task (e.g.: "the", "and", ...).

Following, we apply a filter to eliminate unusual or insignificant words.

We eliminate the swadesh words and apply the lemmatization, the process of grouping together the different inflected forms of a word so they can be analyzed as a single item.

We filter the synsets; the synset represents a concept, and contains a set of Words, each of which has a sense that names that concept (and each of which is therefore synonymous with the other words in the Synset).

Finally, we remove adjective and verb forms.

As final result, we obtain a key term list who, through the semantics of its terms can summarize the concept expressed in the analyzed post. Less detail, i.e. a greater granularity of the information (more posts, whole discussion or whole forum) is easily available by recursively applying described reduction process.

So, the main technologies the system uses are: Jena, a open source Java framework for building Semantic Web applications, used to manage RDF files containing information about users and posted messaged (user, has_created, message) and Java Wordnet Libraries (JWNL) to use Wordnet.

System Interface

The role of Web crawler is to collect Web pages from Internet. Here, the traditional spider detecting algorithm is employed with two parameters as the seed URL and the depth for crawling.

The system interface is showed in following Figure, where the numbers identify each section.

In particular, the TextBox 1 is used to specify forum seed URL. The Button 2 executes IR algorithm to retry keywords. The message viewer 3 shows the messages of the selected topic, the user name, the most relevant words and their occurrences. The keyword list 4 reports the most important extracted keywords; clicking one of the keyword the message viewer will show only the messages containing the selected word. The check boxes 5 allow to enable or disable filters during the keyword extraction algorithm execution.

Following, we show some example of system usage using different forum URL and different contexts (quarterback NFL, IPhone 4G and Xbox).

CONCLUSION AND FUTURE IMPROVEMENTS

Summarization can be evaluated using intrinsic or extrinsic measures; while the first one methods attempt to measure summary quality using human evaluation, extrinsic methods measure the same through a task-based performance measure such the information retrieval-oriented task. In our experiments we utilized intrinsic approach analyzing W3Schools forums, official forum of the W3C (http://www.w3cforum.com/).

We have performed a lot of intrinsic experimental tests obtaining and elaborating a corpus of about 100 threads. This experiment is to evaluate the usefulness of concept extraction in summarization process, by manually reading whole thread content and comparing with automatic extracted concepts. The results show that automatic concept-based summarization greatly improves the performance and produces useful information extraction supporting tutors and making learning communities effective. The extracted concepts represent a good summarization of thread contents.

Advanced in concept-based representation appears as a promising technology for implementing distance learning environment, enabling the orga-

Figure 2.

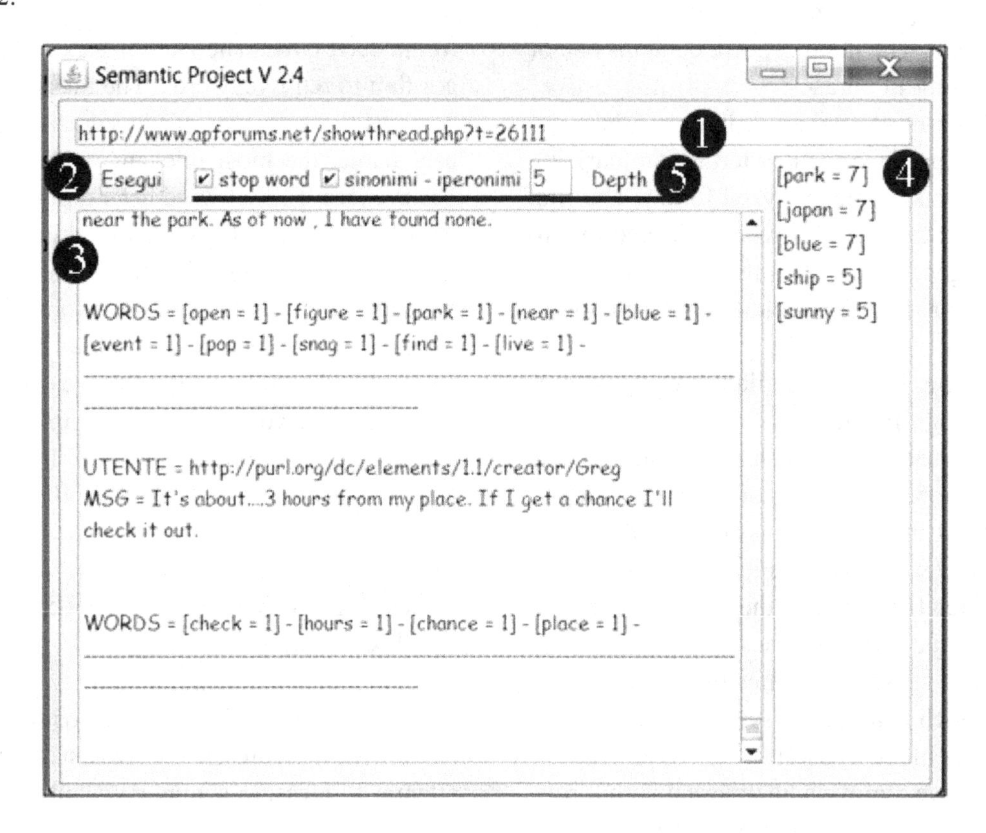

Figure 3.

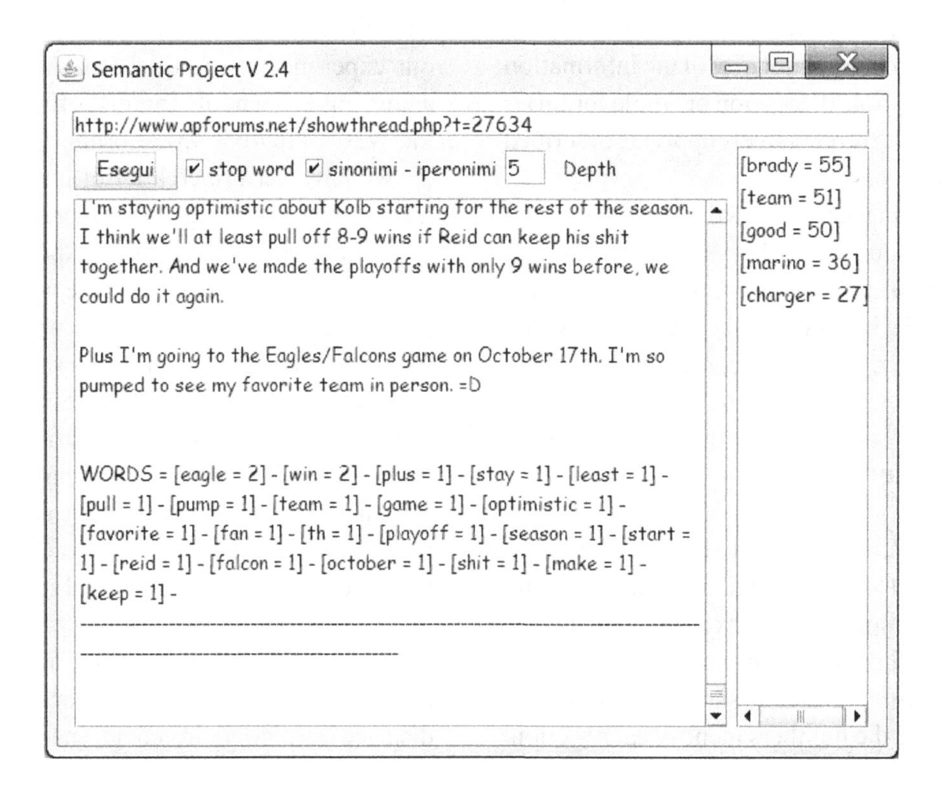

Figure 4.

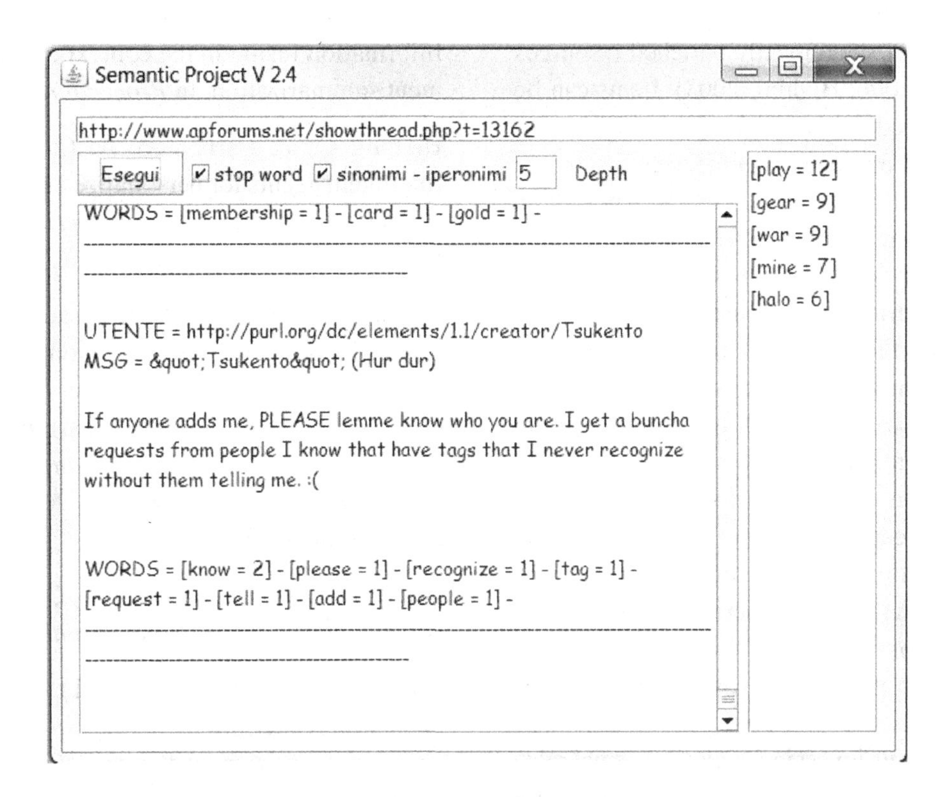

Figure 5.

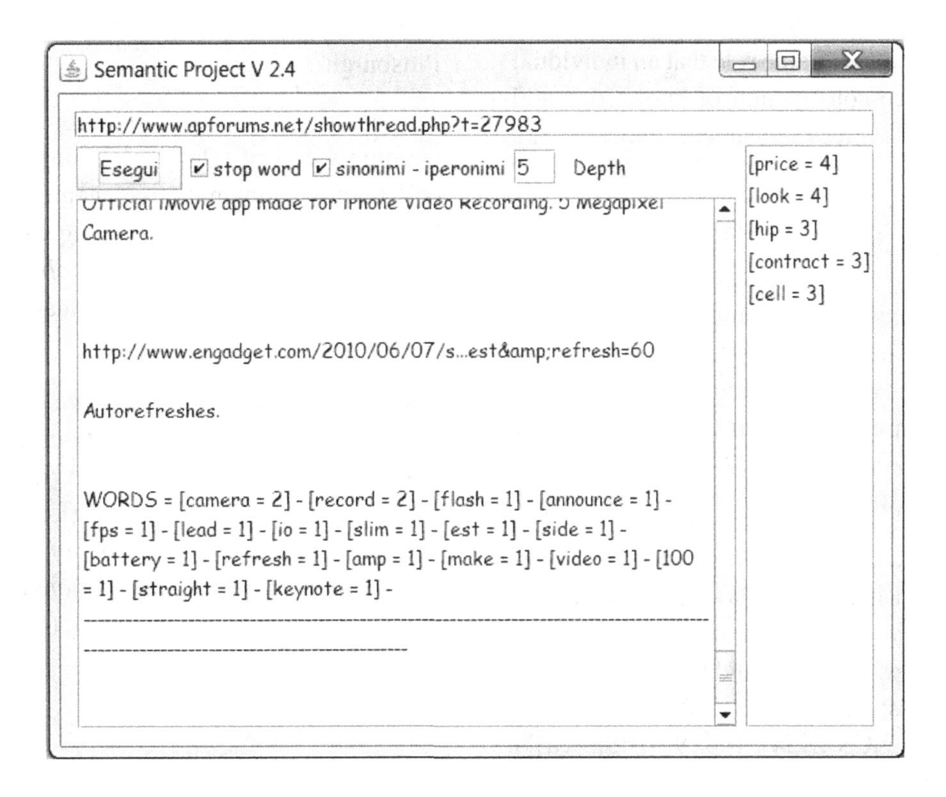

nization and delivery of learning materials around small pieces of semantically enriched resources (Carbonaro, 2006; Bighini, 2003). Items can be easily organized into customized learning courses and delivered on demand to the user, according to her/his profile and business needs (Andronico, 2003; Carbonaro, 2005).

In our experience, concept-based summarization has proven a potentially useful tool to provide a good support for tutors in virtual learning communities. To the best of our knowledge, no systems use concept-based approach to represent online forum information in a learning environment.

We plan to investigate the effectiveness of integrate concept-based approaches such those presented in this chapter within social network analysis (SNA) systems describing relationships and interactions occurring among students and staff participating in computer-supported collaborative learning (Reffay, 2002). In the education context, students invest in developing social relations in order to progress their individual academic goals. The identification of the position a student occupies within the established social network may inform practitioners of the role that an individual plays in the co-construction of knowledge and the types of resources and support they require.

REFERENCES

Aho, A., Chang, S., McKeown, K., Radev, D., Smith, J., & Zaman, K. (1997). Columbia Digital news project: An environment for briefing and search over multimedia. *International Journal on Digital Libraries*, *1*(4), 377–385. doi:10.1007/s007990050030

Andronico, A., Carbonaro, A., Colazzo, L., Molinari, A., & Ronchetti, M. (2003). *Designing models and services for learning management systems in mobile settings*. In Mobile and Ubiquitous Information Access: Mobile HCI 2003 International Workshop, (LNCS 2954), (pp. 90-106). ISBN 3-540-21003-2

Barzilay, R., McKeown, K., & Elhadad, M. (1999). Information fusion in the context of multi-document summarization. In *Proceedings of ACL'99*.

Bighini, C., & Carbonaro, A. (2004). InLinx: Intelligent agents for personalized classification, sharing and recommendation. *International Journal of Computational Intelligence*, *2*(1).

Bighini, C., Carbonaro, A., & Casadei, G. (2003). Inlinx for document classification, sharing and recommendation. In V. Devedzic, J. M. Spector, D. G. Sampson, & Kinshuk (Eds.), *Proceedings of the 3rd Int'l. Conf. on Advanced Learning Technologies*, (pp. 91–95). Los Alamitos, CA: IEEE Computer Society.

Brunn, M., Chali, Y., & Pinchak, C. (2001). *Text summarization using lexical chains.*

Budanitsky, A., & Hirst, G. (2001). *Semantic distance in Wordnet: An experimental, application-oriented evaluation of five measures*. In Workshop on WordNet and Other Lexical Resources. Second meeting of the North American Chapter of the Association for Computational Linguistics, Pittsburgh.

Carbonaro, A. (2006). Defining personalized learning views of relevant learning objects in a collaborative bookmark management system. In Ma, Z. (Ed.), *Web-based intelligent e-learning systems: Technologies and applications* (pp. 139–155). Hershey, PA: Information Science Publishing.

Carbonaro, A. (2010). *Towards an automatic forum summarization to support tutoring*. In M. D. Lytras, P. Ordonez de Pablos, D. Avison, J. Sipior, Q. Jin, W. Leal…D. G. Horner (Eds.), *Technology enhanced learning: Quality of teaching and educational reform, communications in computer and Information Science*, vol. 73. (pp. 141-147). ISBN: 978-3-642-13165-3

Carbonaro, A., & Ferrini, R. (2005). *Considering semantic abilities to improve a Web-based distance learning system*. ACM International Workshop on Combining Intelligent and Adaptive Hypermedia Methods/Techniques in Web-based Education Systems.

Cunningham, H., Maynard, D., Bontcheva, K., & Tablan, V. (2002). GATE: A framework and graphical development environment for robust NLP tools and applications. In *Proceedings 40th Anniversary Meeting of the Association for Computational Linguistics* (ACL 2002). Budapest.

Farell, R. (2002). Summarizing electronic discourse. *International Journal of Intelligent Systems in Accounting Finance & Management*, *11*, 23–38. doi:10.1002/isaf.211

Fellbaum, C. (Ed.). (1998). *WordNet: An electronic lexical database*. Cambridge, MA: MIT Press.

Gunawardena, C. N., & Zittle, F. J. (1997). Social presence as a predictor of satisfaction within a computer mediated conferencing environment. *American Journal of Distance Education*, *11*(3), 8–26. doi:10.1080/08923649709526970

Kolbitsch, J., & Maurer, H. (2006). The transformation of the Web: How emerging communities shape the information we consume. *Journal of Universal Computer Science*, *12*(2), 187–213.

Leacock, C., & Chodorow, M. (1998). *Combining local context and WordNet similarity for word sense identification* (pp. 265–283).

McKeown, K., Barzilay, R., Evans, D., Hatzivassiloglou, V., Kan, M., Schiffman, B., & Teufel, S. (2001). *Columbia multi-document summarization: Approach and evaluation*. Workshop on Text Summarization.

NIST. (2010). *Website on summarization*. Retrieved from http://wwwnlpirnist.gov/projects/duc/pubs.html

Rahman, A., Alam, H., Hartono, R., & Ariyoshi, K. (2001). *Automatic summarization of Web content to smaller display devices*. 6th International Conference on Document Analysis and Recognition, ICDAR01, (pp. 1064-1068).

Reffay, C., & Chanier, T. (2002). *Social network analysis used for modelling collaboration in distance learning groups*. (LNCS 2363), (pp. 31-40).

Resnik, P. (1995). *Disambiguating noun groupings with respect to WordNet senses*. Chelmsford, MA: Sun Microsystems Laboratories.

Rovai, A. P., & Barnum, K. T. (2003). Online course effectiveness: An analysis of student interactions and perceptions of learning. *Journal of Distance Education*, *18*, 57–73.

Shahar, Y., & Cheng, C. (1998). *Knowledge-based visualization of time oriented clinical data*. AMIA Annual Fall Symposium, (pp. 155-9).

Wagner, E. D. (1994). In support of a functional definition of interaction. *American Journal of Distance Education*, *8*(2), 6–26. doi:10.1080/08923649409526852

Yong Rui, Y., Gupta, A., & Acero, A. (2000). Automatically extracting highlights for TV baseball programs. *ACM Multimedia*, 105-115.

Zhou, L., & Hovy, E. (2006). *On the summarization of dynamically introduced information: Online discussions and blogs*. In AAAI Spring Symposium on Computational Approaches to Analysing Weblogs.

Zirkin, B., & Sumler, D. (1995). Interactive or non-interactive? That is the question! An annotated bibliography. *Journal of Distance Education*, *10*(1), 95–112.

Chapter 4
Social Networking and Trust:
Is Personalisation the Only Defence Technique?

Vladlena Benson
Kingston University, UK

ABSTRACT

Based on the literature review of the theory of trust, this chapter aims to provide an insight into trust formation on social networking sites (SNS). An overview of the current state of cybercrime and known ways of threat mitigation helps shed some light on the reasons why social networks became easy targets for Internet criminals. Increasingly, personalisation is seen as a method for counteracting attacks perpetrated via phishing messages. This chapter aims to look specifically at trust in online social networks and how it influences vulnerability of users towards cybercrime. The chapter poses a question whether personalisation is the silver bullet to combat cyber threats on social networks. Further research directions are discussed.

INTRODUCTION

The notion of *trust* elongates from social context to human computer interactions. Much research (e.g. Beudoin, 2008; Jiang, Jones, and Javie, 2008) now recognises the importance of understanding trust and trust-building in online communication, transactions and systems. Online interactions involve various types of risks and entail presence of trust between communicating parties as well as in the applications used for these interactions (Riegelsberger, Sasse, and McCarthy, 2003). Over the past few years there has been an emergence of a number of online social networking sites, such as MySpace, Facebook, Twitter, LinkedIn, to name but a few. Social networking sites (SNS) provide a straightforward, user-friendly and convenient way to connect and share information with other users online. This explains the growing popularity of such SNS as Facebook, which counts more than

DOI: 10.4018/978-1-61520-921-7.ch004

200 million active users. Over 100 million people made logging into Facebook a daily routine and the amount of content shared through the site reached one billion pieces (including web links, stories, blogs posts, photos, etc.) each week (Facebook Statistics, 2009). Features for customising personal profile and privacy settings, peer based rating system and a sense of a secure environment for sharing personal information and content made SNS immensely popular.

The nature of social interactions online forms the basis for trust building and trust-transfer between users of social networks. Online interactions are possible only when not only users trust each other, but also when they have enough faith in the systems they use to transact as well as the organisations which provide them (Riegelsberger et al., 2005). Trust in e-commerce application has attracted a significant research attention (Golbeck, 2008), the concept of trust has been linked to security and used in context of privacy, identity and authorisation. Research efforts have been directed at establishing factors influencing trust in online applications. For example, Dutton and Shepherd (2006) argue that trust on the Internet has been based on two indicators: net-confidence and net-risk. Perceptions of confidence in the Internet technology has been expressed through the level of reliability of information on the Internet, trust in the institutions running the Internet and the confidence in people with whom they conduct online communication and transactions. Net-risk was defined as the perceived exposure to risks or user perceived vulnerability to threats while using the Internet.

Earlier studies showed that those people who use the Internet more tend to gain more trust in technology. Dutton and Shepherd (2006) highlighted that education and experience determine formation of trust among Internet users. They also showed that reinforcement of the digital divides, including life stage factors, social gaps and proximity to the Internet, translates into significant differences in levels of trust developed by people. Based on these arguments and data reported in the 2007 OxIS Survey (Dutton and Helsper, 2008) it is possible infer that higher levels of trust are developed by individuals aged between 18 and 24 that intensively use social media and have the closest proximity to technology. Monitoring the threats from Internet criminals has been the centre of attention of major information security solution providers, e.g. Symantec, MessageLabs, SANs and others. However, research literature demonstrates a lack of sustained research into how social networking users are affected by cybercrime and how they respond to the existing threats. This chapter helps fill the gap in the literature by a) providing an analysis of factors influencing formation of trust in technology in online social networks, b) identifying factors which make social networking users vulnerable to cybercrime and c) provoking a discussion on the reliability of personalisation as the mitigation instrument against social networking vulnerabilities.

The structure of the rest of the chapter is organised as follows: Section 2 provides an overview of the trust theory further linked to the discussion of trends in cybercrime in section 3. Section 4 describes scenarios of Internet attacks perpetrated through social networking sites. Section 5 summarises the factors which influence formation of trust among users of online social networks and examines the aspects of behaviour which make their users prone to cybercrime. The article concludes with the discussion about whether personalisation is the silver bullet to combat cyber threats on social networks. Further research directions are discussed.

Critical Concepts

Social Exchange Theory provides a framework defining trust and helps explain formation of trust in interpersonal and exchange relationships. A general definition can be given to trust as a "*psychological state comprising the intentions to accept vulnerability based upon positive expecta-*

tions of the intentions or behaviour of another" (Rousseau et al., 1998, p.395). Sociological connotations of trust commonly comprise of two elements: confidence and willingness to act based on this confidence (Stompka, 1999). Generally, social trust relationships are viewed as positive in social and collaborative web technology. This is the case for user-generated content, where social trust helps users wade through overwhelming amounts of information, often deceiving, inaccurate and of questionable origin. Social trust helps users find trusted information resources and identify useful content (Ziegel and Golbeck, 2006). Similarity of opinions has been widely used as practical metrics of establishing a trustworthy source of content on the web. However, the basic component of social trust, i.e. willingness to act upon belief, is what makes this type of trust unsuitable in the context of privacy, confidentiality and secure transactions. In fact, in online communication and exchanges, generally termed as transactions, trust makes transaction parties prone to malicious acts. Although many definitions of online trust exist, most of them include the element of vulnerability. For instance, online trust is viewed as *"...willingness to accept vulnerability in an online transaction based on positive expectations regarding future behaviours..."* of transaction parties (Kimery and McCord, 2002). Metrics of interpersonal trust, *"the lubricant of the inevitable frictions of social life"* (Putnam, 2000, p.235), has been expressed and measured as trusting others. Studies measured interpersonal trust through expectations of behaviour of others in certain community situations or with questions about others trying to take advantage of you, people being helpful and being trustworthy (Brehm and Rahn, 1997). Beaudoin (2008) concluded that the greater the Internet use, the greater the interpersonal trust. This conclusion supports earlier findings that trust becomes a function of experience of technology. Dutton and Shepherd (2006) in a survey of the UK population use of new media found that trust on the Internet (its technology, resources and applications) is higher among more experienced users. The authors established that the more time people spend on the Internet, the more trustful they become.

Golbeck (2008) addressed six factors which encourage users to develop social trust. The first two are specific to experience and concur with (Corritore, Kracher, and Wiedenbeck, 2003), the last four are related to reputation in a sense that they reflect influence of opinions from others. The following factors are identified that encourage development of social trust:

- **Prior lifetime of history and events affecting psychological factors.** Although generally related to previous experience, this factor is relevant to past events which affected psychology of a user in a positive or negative way.
- **Past experience with a person and with their friends.** Users are likely to be more trustful if they had positive social experiences.
- **Rumour.** Although not always trustworthy source of information, rumours can influence formation of social trust. If a person is aware of a negative rumour about someone or something, that person is less likely to trust the other.
- **Opinions of the actions undertaken by a person.** Having knowledge of positive actions, which characterise someone, encourages social trust in others.
- **Influence by other people's beliefs.** Negative judgements about a person by other people are likely to discourage trust in that individual and visa versa.
- **Extending trust to gain something.** If there is an incentive or interest to gain an advantage people are more likely to trust others.

Finding a way to determine how much one social networking user can trust the other represents a significant challenge to application of social trust (Golbeck, 2008). Hundreds of social network users are unlikely to form interpersonal trust based on face to face contact. Instead, trust transfer process from one trusted user to a new entity occurs. Estimating trust between users of social networks has been a popular discussion topic in recent literature (e.g. Massa and Avesani, 2007). Algorithms based on network structure and similarity measures among users of social networks have been developed (Avesani, Massa and Telia, 2005). Exploring community structure (Girvan and Newman, 2002) and other complex system approaches derived from probabilistic reasoning and artificial intelligence (Mitchell, 2006) are emerging for computation of trust and prediction of its dynamics in social networks. Simpler general applications of trust formation have been widely used on Internet for years. Including recommender systems (i.e. Amazon pioneered suggestions of shopping items based on profiles), rating and review approaches are widely used in e-commerce. Email message filtering based on measurement of trust of sender origin is another example of trust application (Golbeck and Hendler, 2004). Social networks are enabled by peer-to-peer systems of trust building and rely on trustworthy members. The approach of rating member by peers improves the way of identifying trustworthy users but these rating are useless for third party applications as this information cannot be shared (Golbeck, 2007).

Much research, especially in the early days of e-commerce, has been conducted to understand what encourages people to trust websites. A significant attention was placed on understanding trust in e-commerce website as they involve monetary transactions. According to (Corritore, Kracher, and Wiedenbeck, 2003) three factors affecting trust online include: perception of credibility, risk and ease of use. A large study of formation of trust in e-commerce (Wang and Emurian, 2005) had similar results. In a survey of users of e-commerce websites six major factors that encouraged people to think that a website was trustworthy were identified. While the first two features are specific to perceptions of reputation, the last four factors deal with design and implementation of the website. The factors that impact trust in online environments were identified as:

- **Brand:** Outside reputation of a company independent of its website affects users' willingness to trust website content.
- **Seal of Approval:** Trust is fostered through the evidence of third party certification and security measures.
- **Navigation:** Users are less likely to trust a website if they find it difficult to navigate through its pages.
- **Fulfilment:** Users tend to build more trust as a result of successful order processing, or form loose trust as a result of negative order fulfilment.
- **Presentation:** Layout and presentation of a website encourage users to trust a website.
- **Technology:** The technologies used for website development and operation have a significant impact how much users trust a website.

Overall, all of the above factors are related to the three features identified in to (Corritore, Kracher, and Wiedenbeck, 2003). In addition, according to (Golbeck, 2008) trust is developed over time. The more time users spend on a website, the more informed their decision as to trust a website becomes. The work by Beudoin (2008) follows up on these conclusions relating trust and Internet use over time. The study suggests that Internet use, including social networking, is influenced by user motivation to build social resources, ranging from social contacts, connections and to social interactions.

Social Networks and Internet Threats

While the proportion of online social networks-based communication and transactions is steadily rising (e.g. Golbeck, 2008), concerns of online users about the security of information stored and transmitted online continue to deepen. Findings of the Internet in Britain Report (Dutton, Helsper and Gerber, 2009) showed that a growing proportion of UK society go online to reinforce the networks of family and existing social connections, as well as to meet and be introduced to new people. Internet users, particularly those using a range of communication facilities, are likely to see online social communication as enabling them to be more productive at work and to enhance their personal, financial and economic well-being. It is not surprising that over the past several years online social network users have become targets of cybercrime. Attacks on online communities, especially based upon reputation, have been categorised by (Hoffman, Zage and Nita-Rotaru, 2009). Attackers are motivated by selfish or malicious intent. In this survey attackers were classified into the following types:

- **insider** or **outsider** – based on location of the attacker;
- **loner** or a **part of a coalition** – depending on whether an attacker acts as a part of an organised criminal group or alone;
- **active** or **passive** attacker – based on the type of attacks perpetrated.

In context of this chapter we will focus on active malicious attackers and attack strategies since any form of attack perpetrated via or aimed at a social networking site requires interaction with the system. The next section provides an overview of attacks perpetrated using SNS.

Classification of Internet Threats on Social Networks

The underpinning technology of social media driven by peer to peer interactions becomes an easy mechanism for hackers to target large numbers of users. Here we will consider the following types of active attacks perpetrated through social networks bearing in mind that they are not mutually exclusive and multiple threat vectors can be used to perpetrate an attack:

Sybil Attack – The problem of obtaining multiple legitimate accounts and identities at a very low cost have received significant attention (Friedman et al. 2007) enabling malicious entities to become an internal part of a social network. An adversary then can post malicious content, overhear communication between other entities, harvest valuable personal data.

Malware Attacks – Are not only specific to social network but are increasingly low tech to execute by attaching malicious content to sharable multimedia or text. The 'clickjacking' attack on Twitter in 2009 was just one of many examples of these online threats.

Social Engineering Attacks - Social networks are an ideal ground for social engineering, which is an act of luring unsuspected users into disclosing confidential information by exploiting interpersonal trust. Over the past few years we have seen a growing trend of social engineering attacks propagating through social networks targeting end users with a single purpose of financial gain. Tailored to the nature of social networks, such social engineering attacks as Koobface virus and Scareware links on Digg.com have spread through Facebook with considerable security threats attached.

Hijacking Attacks – By using brute force attacks or other means of retrieving or guessing user passwords an adversary can gain access to a legitimate user's account. The reason criminals seek to abuse social networking accounts is that messages from friends have a better chance of being treated as legitimate by other users.

While not exploiting any particular vulnerability in technology most of these threats require very little expertise and are based on pure social engineering. As discussed by a number of authors (e.g. Galagher 2009, Goldman, 2009) because of the trusted nature of social networks, they are

Figure 1. Phishing email exploiting trustful nature of social media

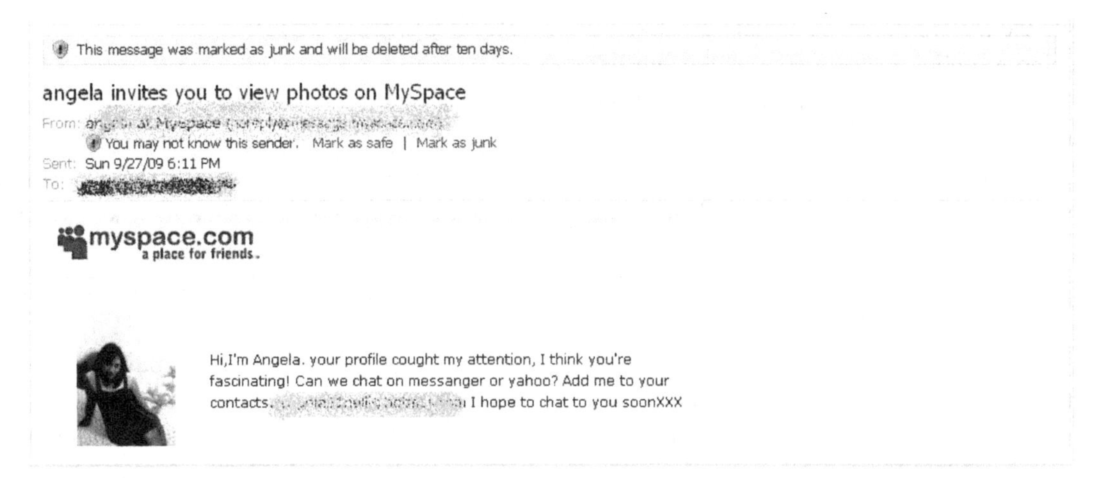

extremely attractive to Internet criminals. Catching user attention by messages such as *'don't click me'*, *'view my video'* or *'look at my profile'*, threats spreading through social media take users to a completely different site, install malware on the victim's computer to steal their personal data or propagate further through the network.

All of the attacks discussed so far are spread by phishing messages which have an appearance of legitimacy when masqueraded as originating through a social network. According to the Internet Threats Report 2008 (Symantec Enterprise Security, 2009), *'Phishing is an attempt by a third party to solicit confidential information from an individual, group, or organization by mimicking (or spoofing) a specific brand, usually one that is well known, often for financial gain'*. Phishing criminals try to hoax Internet users into disclosing sensitive information and use it in identity theft crimes. Phishing usually requires an end user to input their credentials into an online form. Information targeted by phishers includes personal data, such as credit card numbers, online banking credentials, etc.

An example of a phishing email message is shown in figure 1 and may look very similar to a multitude of phishing messages arriving to email boxes world wide. By the time an unsuspected user clicks on the link in the message content the drive-by malware attack already happened and phishing messages to the victim's trusted friends are on the way.

A successful phishing attack may use a number of methods (e.g. Ollman, 2008) to deceive social networking users into revealing their data or do something with their server and web content. The most widespread methods are described below:

Man-in-the-middle – the attacker positions between a legitimate user and the web based social networking site and proxies the communication stream between the systems. This is one of the most successful methods of gaining control of user information and resources through observing and recording transactions between unsuspecting parties.

URL Obfuscation - by means of a phishing message seemingly originating from a trusted social networking site an adversary lures a user into clicking on a malicious hyperlink (URL) leading to an attacker's server. The ways to hide or obscure the destination of a hyperlink are many and are freely available on the web. Some of the common methods are third party's shortened URLs, bad domain names and cross-site scripting. While the in-depth discussion of this type of attack is beyond the scope if this chapter, we will draw the attention to the *Cross-Site Scripting* attack. It exploits the trusted nature of social networks by disguising a malicious web page under familiar URL. For example, a user receives a message from a friend on a social network and follows a URL: http://mysocialnetwork.com/login?URL=http://malicious_site.com/phishing_page.htm. The user thinks that he is directed to his trusted social network. However, by exploiting coding vulnerabilities on the social site, the adversary has managed to refer the user to a falsified login page the attacker's server. The unsuspecting user has no way of distinguishing the malicious URL masqueraded as legitimate and the adversary gains user's login credentials.

Preset Session - is an attack perpetrated by means of a phishing message containing a link to a legitimate application server. In many online applications implementing poor style session management allow user connections to define a SessionID. Users need to be authenticated before proceeding to the restricted page. The adversary's system continuously polls the restricted page of a social network site waiting for the user to click on the phishing message link containing the SessionID predefined by the adversary. The unsuspecting user authenticates and the hacker can carry out an attack.

End-user vulnerabilities are increasing exploited by criminals as web based social networking sites run application on the client side. Social sites run in web browsers and enable multimedia sharing through applications which are constantly under probing for vulnerabilities by Internet criminals. Software manufacturers constantly monitor applications for known vulnerabilities and issue patches and updates for any identified weaknesses. However, unless client side applications are patched and updated regularly, computers of social networking users are open to attacks even if a single unpatched vulnerability remains.

RECOMMENDATIONS AND CONCLUSION

Recent surveys of Internet users in the UK (Dutton and Helsper, 2007; Duton, Helsper and Gerber, 2009) suggest that users have developed a certain degree of tolerance to risks associated with information conveyed through the Internet. It appears that previous experience of internet use and perceived competence in technology are most influential factors in building trust on the Internet (Dutton and Shepherd, 2006). While users are becoming more aware about phishing and other social engineering attacks, social networking sites are being more proactive about scanning downstream sites and isolating malicious content. However, two factors make the quest for safety an ongoing battle: sheer numbers of social networking traffic and reliance on end users to distinguish trusted from malicious content. Many social networking sites have increased their security defences and take technical measures to identify potential phishing attacks. But the shear amount of user generated content makes the task resource intensive and laboursome, therefore SNS rely on customer awareness to identify and report potential offenses. Antiphishing Working Group (2009) puts personalisation of communication as the first line of defence against phishing attacks. They recommend that the personalisation may range from an inclusion of user names, or references to previous communication or some other piece of unique information into messages. Examples include:

"Dear Amily Smith" instead of **"Dear Sir/Madame"** or **"Hi there!"**

Surprisingly enough, social networking user education hardly goes beyond the '*don't click* 'principle. Regular publications of social media safe practices (e.g. VanWyk 2009) reiterate a set of similar recommendations, including:

- **Don't click** on encoded URL's if you are in doubt.
- **Adjust browser settings** (even though browsers are vulnerable to a significant security issues). By allowing active content to run in the browser, users trust third party content to run on their computers.
- **Choose your friends wisely,** i.e. accept connections or content from people you know directly.
- **Avoid running third party applications** on social networking platform, especially new ones.
- **Turn up privacy controls** and limit the number of people who can view one's personal information.
- **Shut down other applications**, i.e. restart the browser and close down other applications before logging to your favourite social networking site.

All of the precautions discussed earlier are certainly valuable and, as suggested by the authors, will help use social media '*in a reasonably safe way*' (Van Wyck, 2009). It seems doubtful that by putting safety of social networking into trustful hands of end users is the only way to mitigate cybercrime exploiting social trust. An alternative solution derives from the experience of the e-commerce sector and their efforts to build consumer trust through third party certification (Jiang, Jones and Javie, 2008). Online social networks are based on social trust and reputation that inform user decisions what information to share with others and deal with information incoming from other users. However, online social networks are enormous and the information needed to decide whether an unknown person or shared content is trustworthy is hardly available. More research attention is needed in understanding, managing and verifying trust on social networks.

The basic component of social trust, i.e. willingness to act upon belief, is what makes this type of trust unsuitable in context of privacy, confidentiality and secure transactions. Whereas social trust helps users find trusted resources online based on profile setting and similarity of opinions (Ziegel and Golbeck, 2006) it makes the nature of social networking prone to cybercrime threats. It is imperative to look for ways in which trust is forged in the ICT development context.

Arguably, future personalisation development for social media must build upon the body of knowledge of secured transactions focusing on confidentiality and integrity principles. Factors encouraging trust from consumers to e-commerce websites, such as seal of approval, branding, navigation, presentation, technology and fulfilment (Corritore, Kracher and Wiedenbeck, 2003) can prove applicable in context of social networking sites. Therefore, more research attention is necessary to address the issues of social trust being a lubricant of content-based transactions on social networks. One of the pressing issues to consider is moving from peer-to-peer formation of trust to third party validation process. However, the challenge of systematically understanding into when trust is needed and works best in online networks still exists. The success of third party trust-based assurance depends on additional research and ICT development efforts in the context of social networks. The article argues for the necessity of further research into the development of entities and identities on social networks. As more and more users, as well as businesses, flock to social networking to interact, communicate and conduct business, it is important to understand the risks and threats involved.

REFERENCES

Adler, P., & Kwon, S. (2002). Social capital: Prospects for a new concept. *Academy of Management Review*, *27*, 17–40. doi:10.2307/4134367

Anti-phishing Working Group. (2009). http://www.antiphishing.org

Avesani, P., Massa, P., & Tiella, R. (2005). A trust-enhanced recommender system application: Moleskiing. *Proceedings of the 2005 ACM Symposium on Applied Computing* (SAC), (pp.1589-1593).

Beudoin, C. E. (2008). Explaining the relationship between Internet use and interpersonal trust: Taking into account motivation and information overload. *Journal of Computer-Mediated Communication, 13*, 550–568. doi:10.1111/j.1083-6101.2008.00410.x

Brehm, J., & Rahn, W. (1997). Individual-level evidence for the causes and consequences of social capital. *American Journal of Political Science, 41*(3), 999–1024. doi:10.2307/2111684

Corritore, C., Kracher, B., & Wiedenbeck, S. (2003). Online trust: Concepts, evolving themes, a model. *International Journal of Human-Computer Studies, 58*(6), 737–758. doi:10.1016/S1071-5819(03)00041-7

Dutton, W. H., & Helsper, E. (2007). *Oxford Internet survey 2007 report: The Internet in Britain*. Oxford, UK: Oxford Internet Institute.

Dutton, W. H., Helsper, E. J., & Gerber, M. M. (2009). *Oxford Internet survey 2009 report: The Internet in Britain*. Oxford Internet Institute, University of Oxford.

Dutton, W. H., & Shepherd, A. (2006). Trust in the Internet as an experience technology. *Information Communication and Society, 9*, 433–451. doi:10.1080/13691180600858606

Ellison, N. B., Steinfield, C., & Lampe, C. (2007). The benefits of Facebook friends: Social capital and college students' use of online social network sites. *Journal of Computer-Mediated Communication*, 12.

Friedman, M., Resnick, P., & Sami, R. (2007). *Algorithmic game theory*. Cambridge, UK: Cambridge University Press.

Galagher, S. (2009). *Social networks a magnet for malware*. IT management e-book. WebMediaBrands.

Girvan, M., & Newman, M. E. J. (2002). Community structure in social and biological networks. *Proceedings of the National Academy of Sciences of the United States of America, 99*, 8271–8276.

Golbeck, J. (2007). The dynamics of Web-based social networks: Membership, relationships, and change. *First Monday, 12*(11).

Golbeck, J. (2008). *Trust on the World Wide Web: A survey*. Hanover, MA: NowPublishers.

Golbeck, J., & Hendler, J. (2004). Reputation network analysis for email filtering. *Proceedings of the First Conference on Email and Anti-Spam*.

Goldman, A. (2009). Businesses lack social media policies. In *IT manager's guide to social networking*. WebMediaBrands.

Hoffman, D. L., & Novak, T. P. (2009). Flow online: Lessons learned and future prospects. *Journal of Interactive Marketing, 23*(1), 23–34. doi:10.1016/j.intmar.2008.10.003

Hoffman, K., Zage, D., & Nita-Rotaru, C. (2009). A survey of attack and defence techniques for reputations systems. *ACM Computing Surveys, 42*(1), 1–16. doi:10.1145/1592451.1592452

Jiang, P., Jones, D., & Javie, S. (2008). How third party certification programs relate to consumer trust in online transactions: An exploratory study. *Psychology and Marketing, 25*(9), 839–858. doi:10.1002/mar.20243

Kimery, K. M., & McCord, M. (2002). Third-party assurances: Mapping the road to trust in e-retailing. *Journal of Information Technology Theory and Application, 4*(2), 64-82.

Lynch, P. D., Robert, J. K., & Srinivasan, S. S. (2001). The global Internet shopper: Evidence from shopping tasks in twelve countries. *Journal of Advertising Research, 41*, 15–23.

Massa, P., & Avesani, P. (2007). Trust metrics on controversial users: Balancing between tyranny of the majority and echo chambers. *International Journal on Semantic Web and Information Systems*, *3*(2).

Mitchell, M. (2006). Complex systems: Network thinking. *Artificial Intelligence*, *170*(18), 1194–1212. doi:10.1016/j.artint.2006.10.002

Naraine, R. (2009). When Web 2.0 becomes security risk 2.0. In *Real business real threats*. Kaspersky Lab.

Ollman, G. (2008). *The phishing guide*. Retrieved from http://logman.tech.officelive.com/Documents/The%20Phishing%20Guide.pdf

Putnam, R. D. (2000). *Bowling alone*. New York, NY: Simon & Schuster.

Riegelsberger, J., Sasse, M. A., & McCarthy, J. D. (2003). The researcher's dilemma: Evaluating trust in computer-mediated communication. *International Journal of Human-Computer Studies*, *58*, 759–781.

Riegelsberger, J., Sasse, M. A., & McCarthy, J. D. (2005). The mechanics of trust: A framework for research and design. *International Journal of Human-Computer Studies*, 62(3), 381-422.

Rousseau, D. M., Sitkin, S. B., Burt, R. S., & Camerer, C. (1998). Not so different after all: A cross-discipline view of trust. *Academy of Management Review*, *23*, 393–404.

Stompka, P. (1999). *Trust*. Cambridge, UK: Cambridge University Press.

Symantec Enterprise Security. (2009). *Internet security threat report* (*Vol. XIV*). Symantec Press.

Urban, G. L., & Sultan, F. (2000). Placing trust at the center of your Internet strategy. *Sloan Management Review*, *42*(1), 39–48.

Van Wyck, K. (2009). How to use Facebook safely. *IT manager's guide to social networking*. WebMediaBrands.

Wang, Y. D., & Emurian, H. (2005). An overview of online trust: Concepts, elements, and Implications. *Computers in Human Behavior*, *21*(1), 105–125.

Yousafzai, S., Pallister, J., & Foxall, G. (2005). E-banking-a matter of trust: Trust-building strategies for electronic banking. *Psychology and Marketing*, *22*(2), 181–201. doi:10.1002/mar.20054

Ziegler, C., & Golbeck, J. (2006). Investigating correlations of trust and interest similarity. *Decision Support Systems*, *43*(2).

Chapter 5
Using Quantum Agent-Based Simulation to Model Social Networks:
An Innovative Interdisciplinary Approach

C. Bisconti
University of Salento, Italy

A. Corallo
University of Salento, Italy

M. De Maggio
University of Salento, Italy

F. Grippa
University of Salento, Italy

S. Totaro
University of Salento, Italy

ABSTRACT

This research aims to apply models extracted from the many-body quantum mechanics to describe social dynamics. It is intended to draw macroscopic characteristics of organizational communities starting from the analysis of microscopic interactions with respect to the node model. In this chapter, the authors intend to give an answer to the following question: which models of the quantum physics are suitable to represent the behaviour and the evolution of business processes? The innovative aspects of the project are related to the application of models and methods of the quantum mechanics to social systems. In order to validate the proposed mathematical model, the authors intend to define an open-source platform able to model nodes and interactions within a network, to visualize the macroscopic results through a digital representation of the social networks.

DOI: 10.4018/978-1-61520-921-7.ch005

INTRODUCTION

In recent years new organizational forms are emerging in response to new environmental forces that call for new organizational and managerial capabilities. Organizational communities are becoming the governance model suitable to build a value-creating organization, representing a viable adaptation to an unstable environment (Clippinger, 1999).

The theoretical framework used in this project to describe the nature and evolution of communities is known as 'complexity science'. According to this approach organizational communities are viewed as "complex adaptive systems" (CAS): they co-evolve with the environment because of the self-organizing behavior of the agents determining fitness landscape of market opportunities and competitive dynamics. A system is complex when equations that describe its progress over time cannot be solved analytically (Pavard & Dugdale, 2000). Understanding complex systems is a challenge faced by different scientific disciplines, from neuroscience and ecology to linguistics and economics.

CAS are called adaptive because their components respond or adapt to events around them (Levin, 2003; Lewin, 1999). They may form structures that somehow maintain their integrity in the face of continuing change. The components of a CAS may follow simple rules and yet produce complex patterns that often change over time.

Organizational communities share many of the characteristics that are used to define a complex adaptive system. Social multipliers, positive feedback, non linearity, evolution, self-organization are phenomena that have been used to explore social interaction in the field of complexity theory.

A number of tools have been developed in recent years to analyse properties and dynamics of complex systems. Amaral and Ottino (2004) identify three types of tools belonging to areas well known to physicists and mathematicians: Social Network theory, Quantum Mechanics, Statistical Physics.

A number of researchers have shed light on some topological aspects of many kinds of social and natural networks (Albert & Barabasi, 2002; Newman, 2003). As a result, we know that the topology of a network is a predictable property of some types of networks that affects their overall dynamic behaviour and explains processes such as: the diffusion of ideas in an organization, the robustness to external attacks for a technological system, the optimization of the relationships among the network components and their effects on knowledge transfer.

Social Network Analysis (SNA) and Dynamic Network Analysis (DNA) are becoming increasingly adopted methodological approaches to study organizational networks during the past years (Wasserman & Faust, 1994; Gloor, 2008).

Social Network Analysis proposes methods and tools to investigate the patterning of relations among social actors. It studies organizational communities, providing a visual representation and relying on the topological properties of the networks so measuring the characteristics of the network object of the analysis.

The focus on the identification of the topological structure of the network based merely on the frequency or intensity of connections represents at the same time an innovation factor and a potential limitation of the approach. The analysis of an organization that does not consider the quality and the content of relations might provide a distorted mirror of the real organizational dynamics.

The main limitation of SNA is to be mainly a quantitative social science method, focused on the structural properties of networks and paying not enough attention to the qualitative issues necessary to understand a phenomenon. Its unit of analysis is not the single actor with its attributes, but the relations between actors, defined identifying the pair of actors and the properties of the relation among them. By focusing mainly on the relations, SNA might underestimate many organizational elements which could influence the ability of an organization to reach its goals. It does not measure how different actors' attributes influence the network configuration. Furthermore, perceptive measures are often ignored by SNA. What seems to be missing in current SNA research is an approach to study how the individual actors' characteristics change the network configuration and performance.

The empirical work on network information advantage is still "content agnostic" (Hansen, 1999). As stated by Goodwin and Emirbayer (1999), SNA globally considered is a framework to investigate the information structure of groups, the structural aspect of relationships, disregarding the content of relationships, and the nodes' properties. Paying attention only to the structural facets of community interactions is like considering all the ties as indistinguishable and homogeneous. In this perspective, actors performing different activities, or involved in different projects, are detected simply as interacting members, with no distinction among individual or group properties that might change over time and influence interactional patterns.

Stinchcombe (1990) proposed a systematic theory of social network coming from the dynamic and causal dimensions of the relationships, to investigate ties among members not only under the quantitative aspect expressed by traditional SNA metrics (e.g. density, centrality, cohesiveness), but also under qualitative considerations related to actors' characteristics and content of ties.

Dynamic Network Analysis uses data coming from the SNA to perform an evolutionary study of the organizational networks, to predict possible network transformation over time. Statistical tools of the DNA are generally suitable for large networks, as they are able to perform multilayer analysis of a network, where nodes of different nature and multiple connections are represented.

In SNA nodes have a static structure, while in DNA nodes have a dynamic nature and evolve time by time taking the role of agents who learn during the simulation as typically happen in this types of model.

The principle that many natural laws come from statistics brought many physicists to apply the models of statistical physics also to the study of behavioural models and to study the dynamics of generation of the organizational networks.

Descriptive and inferential statistics have proven to be of great value to analyse social networks evolution, as they provide a set of suitable tools to summarize key facts about the distribution of actors, attributes, and relations; statistical tools can describe the shape of one distribution, as well as of joint distributions and statistical association. Statistical tools have been particularly helpful in describing, predicting, and testing hypotheses about the relations between network properties.

In recent years, a trend emerged in the application of the statistical physics to several interdisciplinary fields such as biology, information technology, and social sciences. In this new context, physicists showed a growing interest for modeling systems using approaches that might be considered far from their traditional field.

In particular, in social phenomena the basic elements are not particles but individuals, and each of them interacts with a limited amount of peers, who are generally close to them. This number can be considered negligible if compared to the number of people within the system.

The idea of describing social phenomena through physics models is older than the attempt of applying the statistical physics. The discovery of quantum laws in collective properties of a community was revealed for instance by the frequency of births and deaths or by the statistics of crimes. Then the development of statistics brought scientists and philosophers to rely on it to develop a systemic understanding of apparently casual behaviour of individuals. The appearing of new extended databases and the rising of new social phenomena (e.g. the Internet) translated the idea of applying statistical physics to social phenomena into a concrete effort.

Empirical studies indicate also evidences of scalability signals and universal factors (Barabasi and Réka, 1999). Naturally it regards macroscopic phenomena that call for the application of approaches typical of statistical physics for the analysis of social behaviour, to understand for instance high regularity on a large scale and collective effects of the interaction among individuals, considered as relatively simple entities.

If organizational communities are classified as complex systems, their complexity is determined by their size and by the amount of variables they include, that make it hard to describe them in deterministic terms. Complexity also depends on the same actors, who determine the complexity of the structure through their own characteristics. The individual has his own personality regardless of the context in which he or she lives, that could impact on the dynamics of the community which he or she belongs to. Furthermore, it is well known that human behaviour is often unpredictable so that only probabilistic assumptions can be done about the outcome of an action. Although it is not fair to refer human behaviour to a mere quantum phenomenon, some analogies can be used to build a quantum based mathematic structure for the agent that, once his interactions with the external environment are defined, could describe some social phenomena.

In the light of the considerations described above, it seems plausible to build the model of social systems as a many-body system based on a quantum structure. We try for the first time to describe a model of the agent with his own identity regardless of the role and the interactions of the communities he or she belongs to. We also consider determinant in our model the relationship between the state of the agent and the interactions typical of the community in which he or she lives.

Our research relies also on agent-based modeling, whose goal is to address the problem of the connection between the lower (micro) level of the social system and the higher (macro) level.

Agent-based models were primarily used for social systems by Craig Reynolds, who tried to model the nature and characteristics of living biological agents. Reynolds introduced the notion of "individual-based models", which investigates the global consequences of local interactions of members of a population (e.g. plants and animals in ecosystems, vehicles in traffic, people in crowds). In these models individual agents (possibly heterogeneous) interact in a given environment according to procedural rules tuned by characteristic parameters.

Epstein and Axtell introduced in 1996 the first large scale agent model, the Sugarscape, to simulate and explore the role of social phenomena such as seasonal migrations, pollution, sexual reproduction, combat, trade and transmission of disease and culture.

The Artificial Life community has been the first in developing agent-based models (Maes, 1991; Steels, 1995; Varela & Bourgine, 1992), but since then agent-based simulations have become an important tool in other scientific fields and in particular in the study of social systems.

In this context it is worth mentioning the concept of Brownian agent which generalizes the notion of Brownian particle used in statistical mechanics. A Brownian agent is an active particle which possesses internal states, can store energy and information and interacts with other agents through the environment.

Agents interact either directly or in an indirect way through the external environment, which provides feedback about the activities of the other agents. Direct interactions are typically local in time and ruled by the underlying topology of the interaction network. Populations can be homogeneous (i.e., all agents being identical) or heterogeneous. Differently from physical systems, the interactions are usually asymmetrical since the role of the interacting agents can be different both for the actions performed and for the rules to change their internal states.

Agent-based simulations have now acquired a central role in modeling complex systems and a huge literature has been developed about the internal structure of the agents, their activities and the multi-agent features.

OBJECTIVES OF THE RESEARCH PROJECT

This research intend to apply the models used in the quantum mechanics many-body to describe social dynamics. It is designed to draw macroscopic characteristics of the network starting from the analysis of microscopic interactions based on the node model. In the aim to experiment the validity of the proposed mathematic model, the project is intended to define an open-source platform able to model nodes and interactions, to simulate the network behaviour starting from specific defined models, and to visualize the macroscopic results emerging during the analysis and simulation phases through a digital representation of the social network.

Many contributions in the social network analysis field have been focusing on the analysis of social networks of small size, where researchers can be relatively sure about the reliability of the observation related to relational patterns among social actors. Recently, with the rising of online communities, social scientists became more interested to the exploration of social networks based on many nodes, where the interaction model can be analyzed through inferential statistics techniques. The availability of large size networks and the need to conduct the analysis on samples and statistical approximation started a discussion about the reliability of the results based on the observation of small size social networks.

Our research starts with the study of methodologies and tools traditionally used in the field of inferential statistics. Within this perspective, these methodologies are useful to understand the level of reliability of the observed interactions in mirroring the real behaviour of the population under investigation.

A discussion about the limitations and potentialities of different inferential statistical models (e.g. Bayesian Theory) will represent the theoretical basis to suggest new models based on quantum physic principles to model the complex space of the actors and of the interactions among them. Quantum physics is plenty of complex model of interaction between states described by vectors in representative spaces, resulting in well defined collective states. Representative space means the space where the actor (the body of the system) is defined. The actor definition is provided through the definition of his/her "state", which could be expressed as a distribution function named "probability strength" and related to the real actor's probability distribution. This probability strength is a vector of the representative space defined as Hilbert space infinite dimension.

Based on this consideration, our research intends to answer the following question: *Which model or which models from the quantum physics are suitable to model the behaviour and the evolution of organizational communities?*. To answer this question, we propose a framework based on three dimensions:

1. Modeling language and primitives able to describe social phenomena based on the individual characteristics and on the interactions among individuals;
2. Physics and mathematical methods applied to the field of organizational communities analysis;
3. Set of Information Technology tools represented by:
    a. Tools for modeling nodes and networks;
    b. Tools for the network behaviour simulation;
    c. Tools for real and simulated variables analytic visualization.

Figure 1. Operational Framework

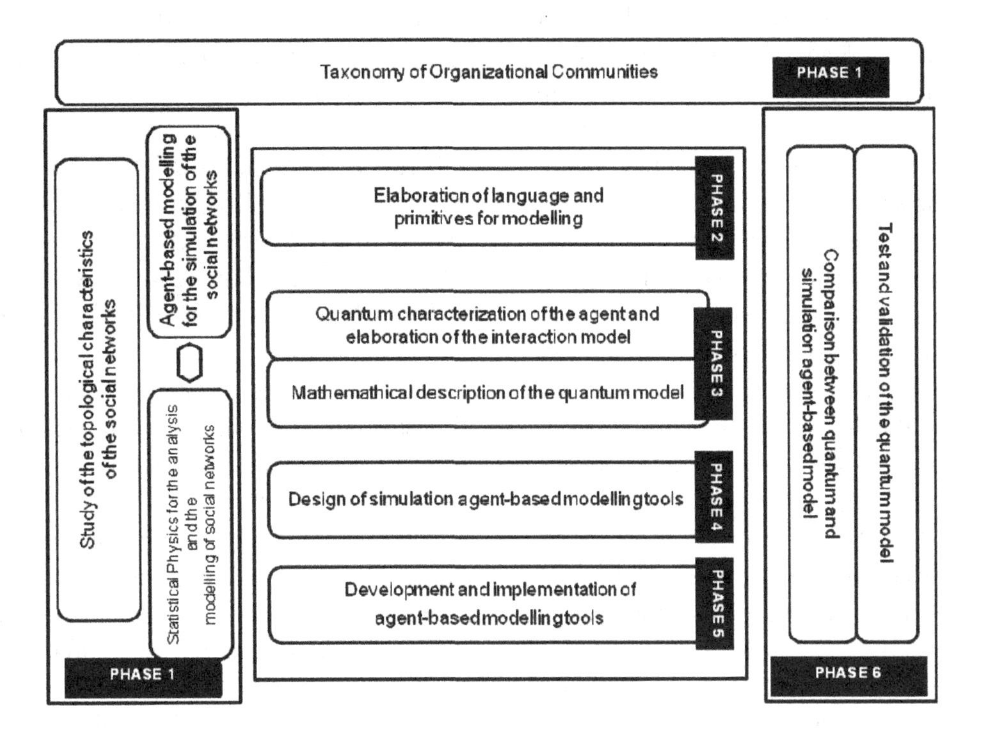

The expected results from the adoption of models traditionally used in quantum mechanics many-body:

1. Definition of the space of configurations and building of the space of operators that afterward will be specified according to their functional form for each particular property that is intended to be described. At this level, we define all the laws that the operators must follow to save the consistency of the model.
2. Development of a formal language to describe at the same time the actors' and the whole network's characteristics, for the effective management of simulation and related results.
3. Development of an actor-actor interaction model able to describe possible communication channels on which an organizational network can be built.
4. Definition of a model of organizational network based on microscopic variables able to describe collective characteristics of the system under investigation.

Figure 1 illustrates the operational framework based on the five phases carried on within this research. The remaining part of the chapter will describe these phases in more details.

Phase 1. State of the Art

The first phase of the project addresses the need of a systematic review on the organizational communities, on social and cultural aspects of communities of practice and on the key factors determining communities' behavior, and is articulated on four steps.

At first, based on the most recent contributions in literature, we will extract the characteristics of the communities operating in an organizational context, defining a taxonomy based on such dimensions as: community purpose, size, degree of physical proximity, leadership and membership, degree of internal diversity, life cycle, sponsorship and degree of institutionalization.

Then the analysis of the communities is integrated with the approaches developed in the field of Social Network Analysis (SNA). In particular, we will consider the contributions in the area of SNA that define structural characteristics of the network such as: density, Group Betweenness Centrality, cohesion. The research conducted in the ONA will help identify the links between structural aspects of the network and the organizational characteristics of the community.

The third step refers to the literature review on statistical approaches for analyzing and interpreting social phenomena, to deepen the knowledge on statistical models and methods for the analysis of relational data, with a particular focus on the relation between actors' behavior and the socio-technical environment in which actors are involved. The applications of descriptive and inferential statistics to study social phenomena will be described, and the main tools able to describe not only the shape of a linear distribution, but also statistical associations and joint distributions. A description of the methodologies and the tools of inferential statistics will be provided with reference to relational data analysis.

The fourth step is about the analysis of the literature on the application of physical models to the study of social dynamics. In addition to the studies on statistical modeling of social behavior, scientific contributions related to their physical modeling will be taken into consideration. This idea of the formulation of physical models to explain events and social dynamics has spread in recent years, bringing an increasing number of scholars to apply their concepts and techniques of physics to other disciplines such as biology, economics and sociology.

Particular attention will be paid to the field of Econophysics, a new science that studies both the behavior of social actors (individuals, groups, organizations), considered as "particles" of a physical system, and the effect of their interactions on the system.

Phase 2. Formulation of a Language and Modeling Primitives to Describe the Social Phenomena on the Basis of Individual Characteristics and Interactions Between Individuals

This phase consists essentially of three steps. The first step is the formulation of a model to describe actor's behavior and the social characteristics that will provide input for defining both the logical structure - defined in the abstract language for modeling actors- and their interactions.

In order to create the quantum mathematical model, it is important to define initially which is the space of configurations where the node has to be modeled, that is to define the properties of the node to which precise functions of distribution will be associated. These functions will act both on ordinary spaces and on spaces with suitably defined characteristics.

This is done in analogy to the description of particles of a quantum system by a wave function that depends both on the coordinates of the ordinary space, and the coordinates of the spin area.

Once defined the space of configurations, we will create the space of operators which will be specified later in their functional form for any particular property that needs to be described. At this level we will present the laws to which operators must obey so as to preserve the consistency of the model.

The models and structures that will shape the organizational community will be described by a formal language that can express at the same time the actors' characteristics and those of the entire network, as well as the management of the simulation and related results.

Although the traditional approach of object-oriented languages will still be considered as a reference point, it will be necessary to study and develop a modeling language easily remapped to XML for its use in the communication between the various software modules that will be implemented in the following tasks.

Phase 3. Formulation of the Mathematical Framework for Building a Quantum Model of the Agent and of the Interactions Between Actors.

This phase addresses the formulation of the mathematical framework for modeling the actor and the interactions with other actors. Following the paradigm of quantum physics, the process consists of building an algebraic abstract structure of operators able to describe the characteristics of the actor. Once identified the algebraic space on which to operate, we will define the possible expressions of the interactions involving actors. This phase has an abstract and mathematical nature, so to consider the whole set of possible properties that need to be represented.

After this phase we formulate the model itself, based on the mathematical framework developed earlier. The stages of formulation of the model follow the typical paradigms of microscopic description of many-body systems.

The actor-agent model connected to this mathematical model is based on the specification of its location along different dimensions, which can be defined in a more or less interdependent way. This opens the way to a vector representation of the state of an agent and a possible tensorial representation of the interactions between agents, and leads to a vision in which the bonds in which the community is structured emerge from such interactions.

A first step to build the model is the definition of microscopic variables. At this stage, based on the algebraic structure built in the previous phases, the microscopic variables describing the state of the actor is described. The so defined configuration spaces are also at the basis of the definitions of the types of possible interactions.

A second step is the formulation of the interactions properties. Once the actor is described, we formulate the basic properties of the interactions. In particular, we take into account interactions of the type "two-bodies", in which actions on multiple configuration spaces are possible (tensorial forces). We also take into account possible three-bodies interactions, though we initially focus only the ones having a "scalar nature". The third step is the construction of a particular model, on which it will be possible to make "ideal" experiments to verify their validity. In the next step of this research, the model will be applied to realistic cases of communities of innovation in the aerospace and defense industry.

Phase 4. Development of Simulation Tools Based on Agent-Based Modeling

The phase of the development of simulation tools based on agent-based modeling is articulated in the three following steps. The first refers to the implementation of the paradigm of actor (node) and community (network). The logical structure for describing the participants and the interactions in algebraic structures, already developed in the previous phases of the project, will be represented in a digital way, using an object-oriented language (known or to be developed), capable of representing actors, communities and simulations in the most efficient, flexible and complete manner. The language will implement polymorphism in addition to the multiple inheritance of classes, in order to create actors with heterogeneous characteristics. This work will also determine the pattern of interaction between actors, and therefore, how the networks will be defined in the simulation environment, in addition to the events to be managed during the temporal evolution of simulation and the interactions made by the graphical interface.

Then the simulation engine based on parallel processing will be designed. The digitalized actors and communities will be simulated in their evolution. This will involve the design of a simulation engine based on parallel processing, that can provide the simulation of large communities, as well as the simulation of several small communities in parallel. To obtain this benefit, it will be important to identify the processes that can be carried out in parallel, so to minimize those to be executed in this sequence.

Finally, the graphic interface will be defined, for handling and interacting with the digital model.

The digital model - described by an object-oriented language - has the parameters of access and setting that will be imported by the graphical objects used in the simulator. These parameters will be graphically set up in a timely manner and in a statistic way. Thus, primitives will be developed able to statistically set the parameters of certain actors within the community. The digital model will also be able to access the output variables for the visual representation of the community's situation, both in the total and in the partial way. To test the network is therefore natural to define the model using graphical tools, by simulating it from a statistical or deterministic configuration, and then by observing the results.

Phase 5. Development and Implementation of Simulation Tools Based on Agents-Based Modeling

The engine designed in the previous phase will be implemented on UNIX-like platform, will be issued with L-GPL license and will be developed in a modular fashion. Among the most important modules to be implemented we include:

- **a module of communication:** communication with the engine will be achieved through a client-server architecture, that will pack in TCP socket XML messages, with which we can prepare and manage the simulation and the results.
- **a module of parallel calculation:** on this module all the parallel operations will be submitted, operations that are related to the simulation and to its evolution over time. This module could be transferred to systems of Grid Computing, if the size of the community to simulate or the number of simultaneous simulations was very large.
- **a module of data storage (both intermediate and final):** the amount of data produced by one or more contemporary simulations will be run by this module, designed to provide high bandwidth in terms of data read/written per second. This modular design will allow the allocation of this part of the engine on a storage area made by many terabytes, so to support simulations of large communities, or of many small communities simulated in parallel.
- **a module of simulated composition of the results:** this module will present two modes of operation. The first mode, aimed at a reporting task. The second mode is interactive, and will make it possible to retrieve data from the storage module and present them in real-time through a graphical representation.

The design of the graphic interface will consist of the implementation of a framework that will implement all the specifications relating to the handling of digital models in a graphical environment. In this framework three software tools will be implemented:

- a tool, that we call *Creator*, through which actors and community will be shaped.
- a tool, that we call *Simulator*, through which, by interacting with the engine it will be possible to set, start and manage the simulation of the community.

- a tool, that we call *Reporter*, which will interact with the engine, and will be able to produce the report of the simulation and analysis of the simulated community.

The modules of the designed open source platform will be validated on samples of data obtained through the involvement of organizational communities. This first test will serve to make a fine-tuning of the applications before the implementation on a large scale.

Phase 6: Validation of the Framework on Cases of Organizational Communities

The first step of this phase is the identification of the organizational communities which will serve as test-bed for the mathematical model. On the basis of evidence extracted from a systematic review of the innovation organization science literature (see Phase 1), we identify the properties of the organizational communities on which the validity of the framework will be tested. These communities will be described in terms of differences in nature, community objective, size, degree of physical proximity, types of leadership and membership, level of internal diversity, life cycle, sponsorship and degree of institutionalization.

The second step to validate the framework is represented by the data collection process. In this phase, the collected data are suitable to be processed through the simulation model developed in Phase 5. We rely on the availability of digital archives (e.g. email, forum threads), which will be complemented by questionnaires designed to obtain a formal representation of individual's skills, roles, personality traits. In particular, we plan to map the following variables:

- information flows: exchanges of email, instant messaging, formal and informal meeting, e-meeting, conference call;
- users profiling: personal characteristics, competencies, professional experience, skills, roles, tenure.

The third step consists in the framework validation through a comparison between simulation data and analysis on existing cases.

The computation of the data collected through the system formulated in Phase 5 provides a representation of the community useful as a yardstick for the configuration of the real community, through the application of the simulation functionalities based on the agent-based modeling provided by the system. The hypotheses - generated by the selection criteria through which communities have been selected – need to be tested based on the result of the comparison between "analysis" and "simulation" performed by the system.

In the fourth and final step of this phase, new digital models are built and analyzed. Objective of the task is to apply the model to real organizational communities and extract possible scenarios in which the framework can be useful for new organizational communities. Based on the results of the first applications carried out in phase 6, managerial recommendations on how to interpret the predictions that emerge from the framework. Figure 2 shows the different activities with their interdependencies.

CONCLUSION

The application of the quantum model and of the simulation model to study the behaviour of organizational communities will support the social network analysis within the description of possible evolutionary scenarios of organizational communities based on the comparison between real and simulated models. The project will provide managerial results related to:

- Definition of guidelines for the application of the analysis and simulation frame-

Figure 2. Representation of the phases and interdependence among activities

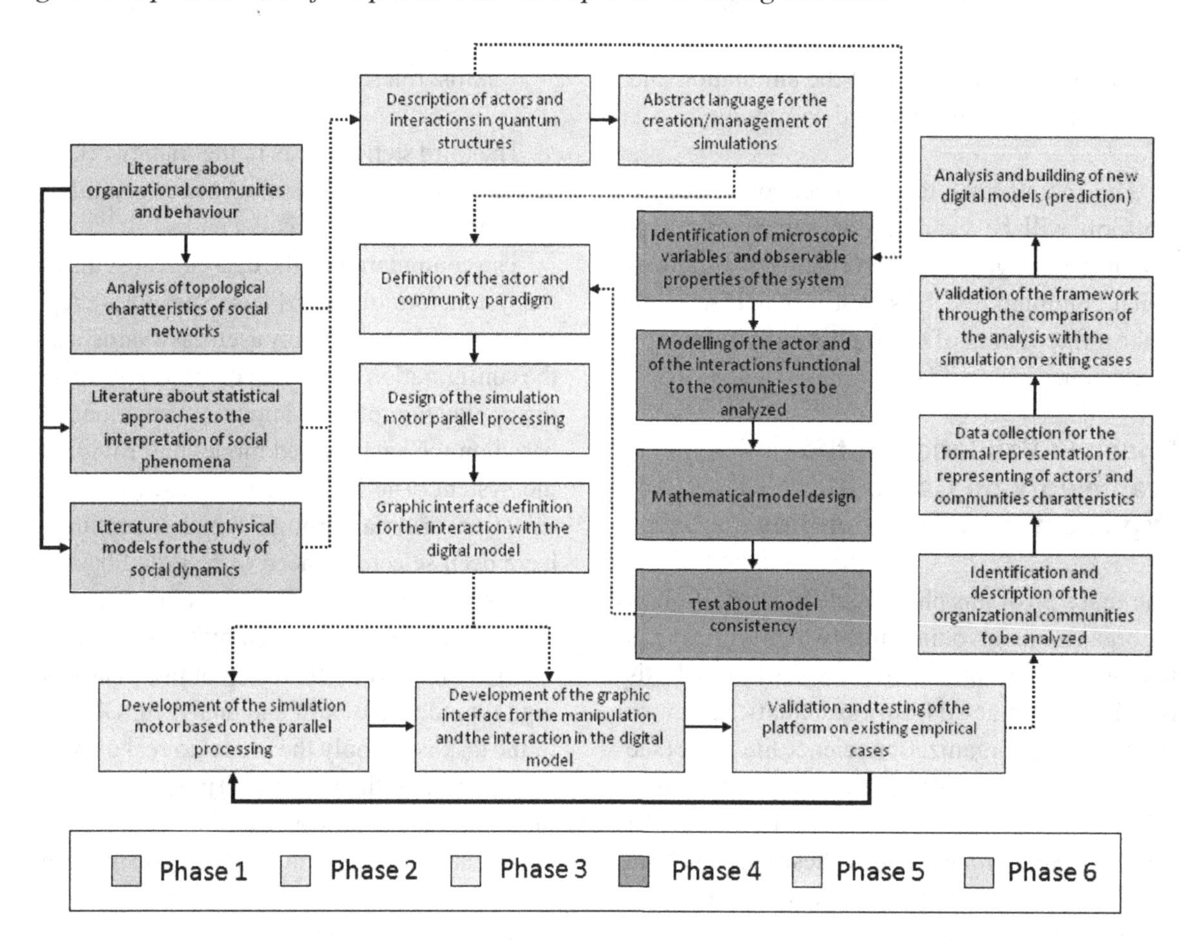

work to organizational communities different in objectives, size, culture, level of institutionalization.

- Development of digital models to be applied to new contexts able to provide analysis and forecasts about the evolution of their organizational properties. Such models will allow to suggest modalities to improve and optimize resources, information flows, competences. The focus of the re-organization could vary according to the choice of the organizational variable to be processed by the analysis and simulation model.

We expect to see the following results from this research:

- Definition of the space of configurations and creation of the space of operators that afterward will be specified according to their functional form for each particular property that is intended to be described.
- Development of a formal language to describe at the same time the actors' and whole network's characteristics.
- Development of an actor-actor interaction model able to describe possible communication channels on which an organizational network can be built.

- Definition of a model of organizational network based on microscopic variables able to describe collective characteristics of the system.

Framework will consist of classes and libraries able to analyze of social networks applying the proposed quantum models. The modules and applications

designed will represent the key instruments applied by scientific communities involved in social network analysis to develop and validate digital models of social phenomena with whom the behaviors of organizational communities will be analyzed and predicted. Then the works obtained will be suitable to be organized and maintained in a simple and effective way. These modules and the applications aim to be interoperable, scalable and to support many simulations at the same time.

To face these requirements the modules of the framework that are intended to be developed will be designed to work on an atomic scale on systems different from an architectural point of view and devoted to support different storage systems located in different places and communicated through TCP/IP networks.

REFERENCES

Albert, M. (2000). *Quantum mechanics*. New York, NY: Dover Publications.

Albert, R., & Barabasi, A.-L. (2002). Statistical mechanics of complex networks. *Reviews of Modern Physics*, *74*, 47–94. doi:10.1103/RevModPhys.74.47

Amaral, L. A. N., & Ottino, J. M. (2004). Complex networks-augmenting the framework for the study of complex systems. *The European Physical Journal*, *38*, 147–162.

Barabasi, A.-L., & Réka, A. (1999). Emergence of scaling in random networks. *Science*, *286*, 509–512. doi:10.1126/science.286.5439.509

Berliant, M., & Fujita, M. (2007). *Knowledge creation as a square dance on the Hilbert cube*. (MPRA Paper No. 4680). Retrieved from http://mpra.ub.uni-muenchen.de/4680/

Clarke, F., & Ekeland, I. (1982). Nonlinear oscillations and boundary-value problems for Hamiltonian systems. *Archive for Rational Mechanics and Analysis*, *78*, 315–333. doi:10.1007/BF00249584

Clippinger, J. H. III. (1999). *The biology of business. Decoding the natural laws of enterprise*. San Francisco, CA: Jossey-Bass Publishers.

Eisert, J., & Wilkens, M. (2000). Quantum games. *Journal of Modern Optics*, *47*, 25–43.

Epstein, J. M., & Axtell, R. (1996). *Growing artificial societies*. Washington, D.C.: Brookings Institution Press.

European Commission. (2005). *The business case for diversity. Good practices in the workplace, Belgium*. Retrieved from http://ec.europa.eu/social/main.jsp?catId=370&langId=en&featuresId=25

Gloor, P., Grippa, F., Kidane, Y. H., Marmier, P., & Von Arb, C. (2008). Location matters-measuring the efficiency of business social networking. *International Journal of Foresight and Innovation Policy*, *4*(3/4), 230–245. doi:10.1504/IJFIP.2008.017578

Goodwin, J., & Emirbayer, M. (1999). Network analysis, culture, and the problem of agency. *American Journal of Sociology*.

Hansen, M. (1999). The search-transfer problem: the role of weak ties in sharing knowledge across organization subunits. *Administrative Science Quarterly*, *44*(1), 82–11. doi:10.2307/2667032

Langton, G. (1995). *Artificial life: An overview*. Cambridge, MA: MIT Press.

Levin, S. A. (2003). Complex adaptive systems: Exploring the known, the unknown and the unknowable. *Bulletin of the American Mathematical Society*, *40*(1), 3–19. doi:10.1090/S0273-0979-02-00965-5

Lewin, R. (1999). *Complexity, life on the edge of chaos* (2nd ed.). Chicago, IL: The University of Chicago Press.

Maes, P. (1991). The agent network architecture (ANA). *SIGART Bulletin*, *2*(4), 115–120. doi:10.1145/122344.122367

Nash, J. F. (1950). Equilibrium points in n-person games. *Proceedings of the National Academy of Sciences of the United States of America*, *36*, 48–49. doi:10.1073/pnas.36.1.48

Newman, M. E. J. (2003). The structure and function of complex networks. *SIAM Review*, *45*(2), 167–256. doi:10.1137/S003614450342480

Pavard, B., & Dugdale, J. (2000). *The contribution of complexity theory to the study of sociotechnical cooperative systems.* Paper presented at the Third International Conference on Complex Systems, Nashua, NH, May 21-26, 2000. Retrieved from http://www-svcict.fr/cotcos/pjs/

Romano, A., De Maggio, M., & Del Vecchio, P. (2009). The emergence of a new managerial mindset. In Romano, A. (Ed.), *Open business innovation leadership. The emergence of the stakeholder university* (pp. 19–65). UK: Palgrave Macmillan.

Steels, L. (1995). When are robots intelligent autonomous agents. *Robotics and Autonomous Systems*.

Stinchcombe, A. L. (1990). *Information and organizations*. Berkley, CA: University of California Press.

Svozil, K., & Wright, R. (2005). Statistical structures underlying quantum mechanics and social science. *International Journal of Theoretical Physics*, *44*(7), 1067–1086.

Varela, F. J., & Bourgine, P. (1992). Toward a practice of autonomous systems. *Proceedings of the First European Conference on Artificial Life*. Cambridge, MA: MIT Press, Bradford Books.

Von Neumann, J., & Mongerstern, O. (1953). *Theory of games and econonmic behavior* (3rd ed.). Princeton, NJ: Princeton University Press.

Wasserman, S., & Faust, K. (1996). *Social network analysis. Methods and applications*. Cambridge, MA: Cambridge University Press.

Chapter 6
An Approach to a Semantic Recommender System for Digital Libraries

José M. Morales-del-Castillo
University of Granada, Spain

Eduardo Peis
University of Granada, Spain

Enrique Herrera-Viedma
University of Granada, Spain

ABSTRACT

One of the key aims of the so-called Information Society is to facilitate the interconnection and communication of sparse groups of people, which can collaborate with each other by exchanging on-line information from distributed sources (Angehrn et al., 2008). In specific contexts, such as in the research and scholarly domain, where many times work is developed relaying on team-based research (Borgman, 2007), finding colleagues and associates to build collaborative relationships has become a crucial matter. Actually, this is one of the pillars of the conduct of research and production of scholarship (Palmer et al., 2009). Nevertheless, this task can be specially difficult when the research activity implies opening new multidisciplinary lines of investigation, since it is hard to know what's hot and who's in in a certain domain out of that of this specialization (even if both areas are related or close to each other).

INTRODUCTION

Due to this, scholarly libraries in general and digital scholarly libraries in particular (which are considered by the research and scholarly community as main nodes to access scientific information) must provide their users new services and tools to ease such kind of tasks.

In this paper we present a filtering and recommender system prototype for digital libraries that serves this community of users. The system makes available different recommender approaches in order to provide users valuable information about resources and researchers pertaining to knowledge

DOI: 10.4018/978-1-61520-921-7.ch006

domains that completely (or partially) fit that of interest of the user. In such a way, users are able to discover implicit social networks where is possible to find colleagues to form a workgroup (even a multidisciplinary one).

OVERVIEW OF THE PROTOTYPE

The system here proposed is based on a previous multi-agent model defined by Herrera-Viedma et al. (2007), which has been improved by the addition of new functionalities and services. In a nutshell, our prototype eases users the access to the information they required by recommending the latest (or more interesting) resources acquired by the digital library, which are represented and characterised by a set of hyperlink lists called *feeds* or *channels* that can be defined using vocabularies such as RSS 1.0 (*RDF Site Summary*) (Beged-Dov et al., 2001). The system is developed by the application of different fuzzy linguistic modeling approaches (both ordinal (Zadeh, 1975) and 2-tuple based fuzzy linguistic modeling (Herera, & Martínez, 2000)) and Semantic Web technologies (Berners-Lee, Hendler, & Lassila, 2001). While fuzzy linguistic modelling (Zadeh, 1975) supplies a set of approximate techniques to deal with qualitative aspects of problems, defining sets of linguistic labels arranged on a total order scale with odd cardinality, Semantic Web technologies allow making Web resources semantically accessible to software agents (Hendler, 2001). In such a way, is possible to improve *user-agent* and *agent-agent* interaction, and settle a semantic framework where software agents can process and exchange information. Besides, the model uses fuzzy linguistic modelling techniques to facilitate the user-system interaction and to allow a higher grade of automation in certain procedures. To increase that grade of automation some techniques of Natural Language Processing are used to create a system thesaurus and other auxiliary tools for the definition of formal representations of information resources. Let's review the main features of all these techniques and technologies.

Semantic Web Technologies

The Semantic Web (Berners-Lee, Hendler, & Lassila, 2001) tries to extend the model of the present Web using a series of standard languages that enable enriching the description of Web resources and make them semantically accessible. To do that, the project is based on two fundamental ideas: i) semantic tagging of resources, so that information can be understood both by humans and computers, and ii) the development of intelligent agents (Hendler, 2001) capable of operating at a semantic level with those resources and infer new knowledge from them (in this way it is possible shifting from keyword search to the retrieval of concepts).

The semantic backbone of the project is the RDF (*Resource Description Framework*) vocabulary (Becket, 2004), that provides a data model to represent, exchange, link, add and reuse structured metadata of distributed information sources and, therefore, make them directly understandable by software agents. RDF structures the information into individual assertions (resource, property, and property value triples) and uniquely characterises resources by means of Uniform Resource Identifiers or URI's, allowing agents to make inferences about them using Web ontologies (Gruber, 1995) (Guarino, 1998) or work with them using simpler semantic structures, like conceptual schemes or thesauri.

As we can see, the Semantic Web basically works with information written in natural language (although structured in a way that can be interpreted by machines). For this reason, it is usually difficult to deal with some problems that require operating with linguistic information that has a certain degree of uncertainty (as, for instance, when quantifying the user's satisfaction in relation to a product or service). A possible solution could be the use of fuzzy linguistic modelling techniques as a tool for improving the *communication* between system and user.

Tuple and Ordinal Fuzzy Linguistic Modelling Approaches

Fuzzy linguistic modelling (Zadeh, 1975) supplies a set of approximate techniques appropriate to deal with qualitative aspects of problems. The ordinal linguistic approach is defined according to a finite set S of linguistic labels arranged on a total order scale and with odd cardinality (7 or 9 tags):

$$\{s_i, i \in H = \{0,\ldots,T\}\}$$

The central term has a value of "approximately 0.5" and the rest of the terms are arranged symmetrically around it. The semantics of each linguistic term is given by the ordered structure of the set of terms, considering that each linguistic term of the pair (s_i, s_{T-i}) is equally informative. Each label s_i is assigned a fuzzy value defined in the interval [0,1], that is described by a linear trapezoidal property function represented by the 4-tupla $(a_i, b_i, \alpha_i, \beta_i)$ (the two first parameters show the interval where the property value is 1.0; the third and fourth parameters show the left and right limits of the distribution). Additionally, we need to define the following properties:

$$1. - \text{The set is ordered} : s_i \geq s_j \quad if \quad i \geq j.$$
$$2. - \text{Negation operator} : Neg(s_i) = s_j, \text{ with } j = T - i.$$
$$3. - \text{Maximization operator} : MAX(s_i, s_j) = s_i \text{ if } s_i \geq s_j.$$
$$4. - \text{Minimization operator} : MIN(s_i, s_j) = s_i \text{ if } s_i \leq s_j.$$

Besides, it is necessary to define aggregation operators, such as the *Linguistic Ordered Weighted Averaging* (LOWA) operator (Herrera, Herrera-Viedma, & Verdegay, 1996), which are capable to combine linguistic information.

To develop our model we have also applied another approach to model the linguistic information: the 2-tuple based fuzzy linguistic modelling (Herrera, & Martínez, 2000). This approach allows reducing the information loss usually yielded in the ordinal fuzzy linguistic modelling (since information is represented using a continuous model instead of a discrete one) but keeping its straightforward word processing.

In this context, if we obtain a value $\beta \in [0, g]$ and $\beta \notin \{0,\ldots,g\}$ as a result of a symbolic aggregation of linguistic information, then we can define an approximation function to express the obtained outcome as a value of the set S. The fundamental base of this approach is the concept of "*symbolic translation*" (Herrera, & Martínez, 2000). Let β the result of aggregating the indexes of a linguistic term set S. Given i= *round* (β) and $\alpha = \beta - i$, such that $i \in S[0, g]$ and $\alpha \in [-0.5, 0.5)$, then α is what we call *symbolic translation*, i.e. the difference between the information expressed by β and the nearest linguistic label $s_i \in S$. Therefore, given a linguistic term set $S = \{s_0, s_1, s_2, s_3, s_4, s_5, s_6\}$ and $\beta = 3.3$ as a result of a symbolic aggregation operation, we could represent this value through the linguistic 2-tuple $\Delta(\beta) = (s_3, +0.3)$.

Natural Language Processing Techniques (NLP)

Natural Language Processing (NLP) consists of a series of linguistic techniques, statistic approaches, and machine learning algorithms (mainly clustering techniques) that can be used, for example, to summarize texts in an automatic way, to develop automatic translators, or to create voice recognition software.

Another possible application of NLP would be the semiautomatic construction of thesauri using different techniques. One of them consists in determining the lexical relations between the terms of a text (mainly synonymy, hyponymy and hyperonymy) (Hearst, 1992) and, this way, extracting those terms that are more representative for its specific domain (Aussenac-Gilles, Biébow, & Szulman, 2000). It is possible to elicit these relations making use of linguistic tools (like WordNet (2009)) and clustering techniques. WordNet (Miller, 1995) is a powerful multi-language lexical database where each one of its entries is defined, among other elements, by their synonyms (*synsets*), hyponyms and hyperonyms. As a consequence, once given the most important terms of a domain, WordNet can be used to create from them a thesaurus after leaving out all those terms that have not been identified as belonging or related to the domain of interest (Missikof, Navigli, & Velardi, 2002).

This tool can also be used together with clustering techniques, for example, to group documents of a corpus in a set of nodes or clusters, depending on their similarity. Each of these clusters is described by the most representative terms of their documents. These terms make up the most specific level of a thesaurus, and are used to search in WordNet their synonyms and most general terms (Khan, & Luo, 2002), contributing with the repetition of this procedure to the bottom-up development process of the thesaurus.

Although there are many others, these are some of the most known techniques of semiautomatic thesaurus generation (semiautomatic since, needless to say, it is necessary the supervision of the experts involved in their creation to determine the validity of the final result).

Based on the infrastructure defined by all these technologies, the system is able to filter and recommend resources from two different approaches that are explained in the following section.

RECOMMENDER APPROACHES

Traditionally, filtering and recommender systems have been classified into two categories (Pospescul et al., 2001): systems that provide recommendations about a specific resource according to the opinions given about that resource by different experts with a profile similar to that of the active user (known as collaborative recommender systems) and systems that generate recommendations according to the similarity of a resource with other resources assessed by the active user (i.e. content-based recommender systems).

In both of them, the likeness can be measured using different similarity functions (Salton, 1971) (van Rijsbergen, 1979), which are usually interpreted in a *linear way* (i.e. the higher the similarity value is, the more relevant it is to generate a recommendation). This is what we call *monodisciplinary* approach, since it lets users deepen into their knowledge in a specific area.

Nevertheless, as discussed above, it's quite common (and almost a need) for many researchers the need to keep the track of new developments and advances in other fields related to their specialization domain. In this way, it is possible for them to widen their research scope, open new research lines and create multidisciplinary workgroups.

In such circumstance, users require to get recommendations about resources whose topics are related to (but not exactly fit) their preferences, but without modifying their starting preferences at all. In this case it makes sense considering as relevant an interval of *mid-range* similarity values instead of those close to one (i.e. both extremely similar and dissimilar similarity values are discarded). Therefore, it would be necessary defining some kind of center function (Yager, 2007) (see figure 1) that enable constraint the range of similarity values we are going to consider as relevant. In our model, the interpretation of similarity is defined by a Gaussian function μ as the following:

$$\mu(Sim(p_i, r_j)) = e^{(Sim(p_i, r_j)-k)^2}$$

where *Sim* (p_i, r_j) is the similarity measure among the resources p_i and p_j, and k represents the center value around which similarity is relevant to

Figure 1. Gaussian center function

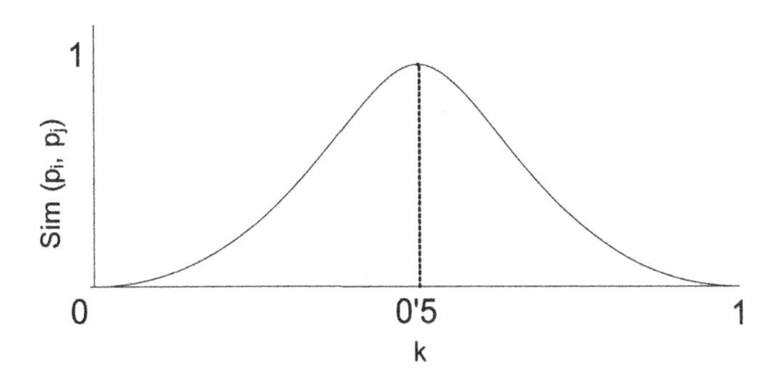

generate a recommendation (in this case *k*=0'5). This is what we call *multidisciplinary* approach.

ARCHITECTURE, ELEMENTS AND MODULES

In this section we describe the architecture, main elements and modules of the system.

Architecture

To carry out the filtering and recommendation process we have defined 3 software agents (interface, task and information agents) that are distributed in a 5 level hierarchical architecture (see figure 2):

- *Level 1. User level*: In this level users interact with the system by defining their preferences, providing feedback to the system, etc.
- *Level 2. Interface level*: This is the level defined to allow interface agent developing its activity as a mediator between users and the task agent. It is also capable to carry out simple filtering operations on behalf of the user.
- *Level 3. Task level*: In this level is where the task agent (normally one per interface agent) carries out the main load of operations performed in the system such as the generation of information alerts or the management of profiles and RSS feeds.
- *Level 4. Information agents level*: Here is where several information agents can access system's repositories, thus playing the role of mediators between information sources and the task agent.
- *Level 5. Resources level*: In this level are included all information sources the system can access such as a full-text documents repository and a set of resources described using RDF-based vocabularies (Beckett, 2004) (RSS feeds containing the items featured by the digital library, a user profile repository and a thesaurus that describes the specialization domain of the library).

The underlying semantics of the different elements that make up the system (i.e. their characteristics and the semantic relations defined among them) are defined through several interoperable web ontologies (Gruber, 1995) (Guarino, 1998) described using the OWL vocabulary (McGuinnes, & van Harmelen, 2004).

The thesaurus, specifically, is a key element since (among other tasks), assists users in the creation of their profile and allows automating the alerts generation. That's the reason why it is critical

Figure 2. Levels of the filtering and recommender system

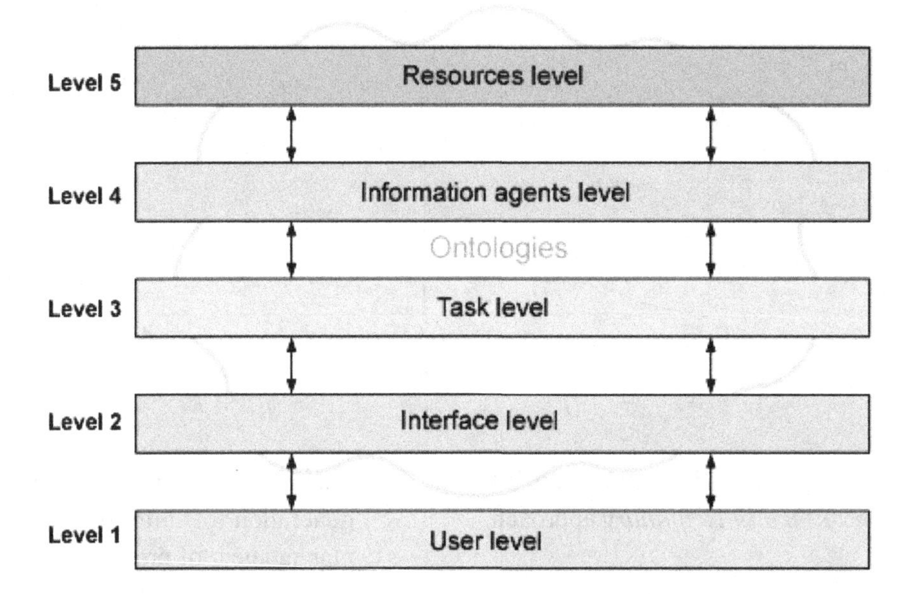

to define the way in which we create this tool. In this work we propose a specific methodology for the semiautomatic development of thesauri using Natural Language Processing techniques (NLP). In the next section we describe a methodology to semi-automatically generate it.

Thesaurus

An essential element of this SDI service is the thesaurus, an extensible tool used in traditional libraries, that allows organizing the most relevant concepts in an specific domain, defining the semantic relations established between them (basically, equivalence, hierarchical and associative relations). The functions define for the thesaurus in our system include helping in the indexing of RSS feeds items, and in the generation of information alerts and recommendations.

To create the thesaurus we follow the methodology suggested by Pedraza-Jiménez, Valverde-Albacete, and Navia-Vázquez (2006).

The learning technique used for the creation of a thesaurus includes four phases: pre-processing of documents, parameterization of the selected terms, conceptualization of their lexical stems, and generation of a lattice or graph that shows the relation between the identified concepts.

Essentially, the aim of the pre-processing phase is to prepare the documents' parameterization, excluding those elements regarded as superfluous. We have developed this phase in three stages: elimination of tags (stripping), standardization and stemming.

In the first stage, all the tags (HTML, XML, etc) that can appear in the collection of documents are eliminated. The second stage is the standardization of the words in the documents in order to facilitate and improve the parameterization process. At this stage, in the first place, the acronyms and N-grams (bigrams and trigrams) that appear in the documents are identified using lists that were created for that purpose.

Once we have detected the acronyms and N-grams, the rest of the text is standardized. Dates and numerical quantities are standardized substituting them by a script that identifies them. Besides, all the terms (except acronyms) are changed to small letters and punctuation marks are removed. Finally, a list of function words is used to eliminate from the texts articles, determiners, auxiliary verbs, conjunctions, prepositions, pronouns, interjections, contractions, and grade adverbs.

All the terms are stemmed to facilitate the search of the final terms and to improve its calculation during the phase of parameterization. To carry out this task, we have used Morphy, the stemming algorithm used by WordNet. This algorithm implements a group of functions that, first of all, allow checking if a term is an exception that does not need to be stemmed, and then, converting words that are not exceptions to their basic lexical form. Those terms that appear in the documents but are not identified by Morphy are eliminated from our experiment.

The parameterization phase has a minimum complexity. Once identified the final terms (roots or bases), they are quantified assigning a weight to each term of each document. Such weight is obtained by the application of the scheme *tf-idf (term frequency- inverse document frequency),* a statistic measure that makes possible the quantification of the importance of a term or N-gram in a document depending on its frequency of appearance in it and in the collection of documents it belongs to.

Finally, once the documents have been parameterized, the associated meanings of each term (lemma) are extracted by searching them in WordNet (specifically, we use WordNet 2.1 for UNIX-like systems). Thus, we get the group of *synsets* associated to each word. The group of hyperonyms and hyponyms are extracted also from the vocabulary of the analyzed collection of documents.

The generation of our thesaurus, id est, the identification of descriptors that better represent the content of documents, and of the underlying relations between them, is achieved using Formal Concept Analysis techniques.

This categorization technique uses the theory of lattices and ordered sets to find abstraction relations from the groups it generates. What is more, this technique allows clustering the documents depending on the terms (and synonyms) it contains. Besides, a lattice graph is generated according to the underlying relations among the terms of the collection, taking into account the hyperonyms and hyponyms extracted. In that graph, each node represents a descriptor (namely, a group of synonym terms) and clusters the set of documents that contain it, being linked such node to those with which it has any relation (of hyponymy or hyperonymy).

Once the thesaurus is obtained by identifying its terms and the underlying relations among them, it can be automatically represented using the SKOS (Simple Knowledge Organization System) vocabulary (Isaac, & Summers, 2008).

This methodology applied to different documental corpora could allow a digital library to obtain different thesauri that can be used to filter and recommend resources to users specialized in different domains.

Modules

In the prototype there are also defined 3 main activity modules:

- *Information push module*: This module is responsible for generating and managing the information alerts to be provided to users. The similarity between user profiles and resources is measured according to the hierarchical lineal operator defined by Oldakowsky and Byzer (2005), which takes into account the position of the concepts to be matched in a taxonomic tree. Once defined this similarity value, the relevance of resources or profiles is calculated according to do the concept of *semantic overlap*. This concept tries to ease the problem of measuring similarity using taxonomic operators, since all the concepts in a taxonomy are related in a certain degree and, therefore, the similarity between two of them would never reach 0 (i.e. we could find relevance values higher than 1 that can hardly be normalized). The underlying idea in this concept is determining areas of maximum semantic intersection between the concepts in the taxonomy. To obtain the relevance of profiles to other profiles we define the following function:

$$Sim(P_i,P_j) = \frac{\sum_{k=1}^{MIN(N,M)} H_k(Sim(\alpha_i,\delta_j))\left(\frac{\omega_i+\omega_j}{2}\right)}{MAX(N,M)}$$

where $H_k(Sim(\alpha_i, \delta_j))$ is a function that extracts the k maximum similarities defined between the preferences of $P_i = \{\alpha_{1,\ldots,}\ \alpha_N\}$ and $P_j = \{\delta_{1,\ldots,}\ \delta_M\}$, and ω_i, ω_j are the corresponding associated weights to α_i and δ_j. When matching profiles $P_i = \{\alpha_{1,\ldots,}\ \alpha_N\}$ and items $R_j = \{\beta_{1,\ldots,}\ \beta_M\}$, since subjects are not weighted, we will take into account only the weights associated to preferences so the function in this case is slightly different:

$$Sim\left(P_i,R_j\right) = \frac{\sum_{k=1}^{MIN(N,M)} H_k\left(Sim\left(\alpha_i,\beta_j\right)\right)\omega_i}{MAX\left(N,M\right)}$$

- *Feedback or user profiles updating module*: In this module, the updating of user profiles is carried out according to users' assessments about the set of resources recommended by the system. This updating process consists in recalculating the weight associated to each preference in a profile and adding new entries to the recommendations log stored in every profile. We have defined a matching function, which rewards those preference values that are present in resources positively assessed by users and penalized them, on the contrary, when this assessment is negative. Let $e_j \in S'$ be the satisfaction degree provided by the user, and $\omega^j_{il} \in S$ the weight of property i (in this case i=«Preference») with value l. Then, we define the following updating function g: S'x S→S:

$$g(e_j,\omega^j_{li}) = \begin{cases} s_{Min(a+\beta,T)} & if\ s_a \le s_b \\ s_{Max(0,a-\beta)} & if\ s_a > s_b \end{cases}$$

$$s_a, s_b \in S \mid a,b \in H = \{0,\ldots T\}$$

where, (i) $s_a = \omega^j_{li}$; (ii) $s_b = e_j$; (iii) a and b are the indexes of the linguistic labels which value ranges from 0 to T (being T the number of labels of the set S minus one), and (iv) β is a bonus value which rewards or penalize the weights of the preferences. It is defined as β=*round* (2|b-a|/T) where *round* is the typical round function.

- *Collaborative recommendation module*: The aim of this module is generating recommendations about a specific resource in base to the assessments provided by different experts with a profile similar to that of the active user. The different recommendations (expressed through linguistic labels) are aggregated using the *Linguistic Ordered Weighted Averaging* (LOWA) operator (Herrera, Herrera-Viedma, & Verdegay, 1996), which is capable to combine linguistic information. It also allows users to explicitly know the identity and institutional affiliation data of these experts in order to contact them for any scholarly purpose. This feature of the system implies a total commitment between the digital library and its users since their altruistic collaboration can only be achieved by granting that their data will exclusively be used for contacting other researchers subscribed to the library. Therefore, becomes a critical issue defining privacy policies to protect those individuals that prefer to be *invisible* for the rest of users. Nevertheless, we have to point out that this functionality is still in development and has not been implemented yet.

OPERATIONAL EXAMPLE

To clarify the performance of the system here we show an operational example. Let's start defining a set of premises:

- A generic user that wants to obtain *monodisciplinar* recommendations from the system, with a profile *P* where preferences α_1, α_2 (N=2) and their associated weights ω_1, ω_2 are defined.
- An item *R* of the RSS feed of the system represented by the subjects β_1, β_2, β_3 (M=3).

First of all the system proceeds to calculate the similarity between the resources in the RSS feed and the profile of the active user applying the taxonomic linear operator defined by Oldakowsky and Byzer (2005). Let α_1 be the concept "*Control instruments*" with a depth of 2 in the thesaurus of the system and β_2 the concept "*Record group classification*" with a depth of 3, being 6 the maximum depth of the thesaurus (see figure 3). As the common parent (*ccp*) of both concepts is "*Archival Science*" (which depth is 0 by default), the distance between them is $d(\alpha_1, \beta_2)= 0.83$, and its associate similarity $Sim(\alpha_1, \beta_2)= 0.17$.

The rest of distances and corresponding similarities are respectively shown on tables 1 and 2:

In the next step, the relevance of the item R to the profile P is calculated. Let the importance value for the preference α_1 be the linguistic label "*Very high*" (i.e. ω_1=0.83) and for α_2 the label "*Medium*" (i.e. ω_2= 0.5). Besides, if the number of preferences and subjects is respectively N=2 and M=3, then the 3 maximum similarities are chosen to calculate the relevance value (in this case, Sim (α_1, β_3)=0.88, Sim (α_2, β_1)=0.84, and Sim (α_2, β_2)=0.93). The resulting relevance value is *Rel (P, R)*= 0.54 so, as the relevance threshold has been fixed in *k*=0.50, the resource R is selected to be retrieved.

Then, applying the 2-tuple based fuzzy linguistic modeling approach (Herrera, & Martínez, 2000), relevance is displayed as a linguistic label

Table 1. Distances between preferences and subject concepts

Preferences/Subjects	β1	β2	β3
α1	0.21	0.83	0.12
α2	0.16	0.07	0.35

Table 2. Similarities between preferences and subject concepts

Preferences/Subjects	β1	β2	β3
α1	0.79	0.17	0.88
α2	0.84	0.93	0.65

Figure 3. Sample concepts in the thesaurus

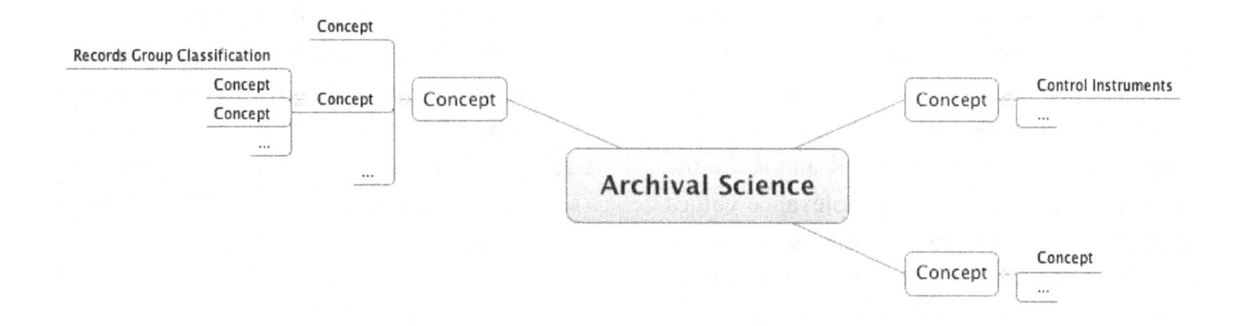

extracted from the linguistic variable "*Relevance level*" together with a numeric value (also called "*symbolic translation*"). Therefore, for the relevance value *Rel (P, R)*= 0.54, the outcome is "*Medium* + 0.04". This implies that "*Medium*" is the closest linguistic label for that relevance value, and that the corresponding numeric value for this label has been exceeded in 0.04.

The following step consists in searching profiles (similar to the profile of the active user) with recommendations about the resource R in order to generate a collaborative recommendation. Supposed two users that have respectively assessed the resource R with the linguistic labels "*High*" and "*Medium*" (which have been extracted from the linguistic variable "*Level of satisfaction*"), when applying the LOWA operator (Herrera, Herrera-Viedma, & Verdegay, 1996) the resulting aggregated label is the following: k= MIN{6,3 + round (0.4*(4-3))}=3. Then l_k= "*Medium*".

As the non-weighted average similarity of the preference α_1 (with a value of 0.80) is lower than that of α_2 (with a value of 0.88), this last preference value will be the chosen to be updated. Let's see an example of the updating process.

Supposed the user assesses the resource R (which has satisfied his information needs) defining a satisfaction level with the linguistic label e_j="*Very High*" (where $e_j \in S'$= *{null, very low, low, medium, high, very high, total}*). In this case, the associated weight to α_2 is $\omega^j_{(Preference,\ \alpha 2)}$= "*Medium*"(where $\omega^j_{li} \in S$ = *{null, very low, low, medium, high, very high, total}*). Considering that $s_a s_b$, whose index values are *a*=3 and *b*=5, and *T*=6, we have that β=*1*, so the new associated weight for α_2 is increased in a factor of one $(\omega^j_{(Preference,\ \alpha 2)})'$= *g* (*Very high, Medium*) = "*High*".

If the user decides to get multidisciplinary recommendations the process is carried out in a slightly different manner. Let R and R' be the set of retrieved resources with relevance values *Rel(P, R)*=0.57 and *Rel(P, R')*=0.83 respectively the system recalculates both relevance values according to the centering function: μ(Rel (P, R))=1.005; μ(Rel (P, R'))=1.110. Then, the system re-arranges the retrieved items and considers as more relevant the values which are closer to one (in this case, R is more relevant than R').

PROTOTYPE EVALUATION

We have set up an experiment to evaluate the content-based module of the prototype in terms of precision (Cao, & Li, 2007) and recall (Cleverdon, Mills, & Keen, 1966) (since the collaborative recommendation module is not fully implemented yet and suffers from *cold start problem* (Schein, Popescul, & Ungar, 2002). These two measures (together with the F1 measure (Sarwar et al., 2000) are usually used in filtering and recommender systems to assess the quality of the set of retrieved resources.

To carry out the evaluation and according to users' information needs, the set of items recommended by the system have been classified into four basic categories: relevant suggested items (Nrs), relevant non-suggested items (Nrn), irrelevant suggested items (Nis) and irrelevant non-suggested items (Nin). We have also defined other categories to represent the sum of selected items (Ns), non-selected items (Nn), relevant items (Nr), irrelevant items (Ni), and the whole set of items (N).

Based on to these categories we have defined in our experiment precision, recall and F1 as follows:

Precision: Ratio of selected relevant items to selected items, i.e., the probability of a selected item to be relevant. P= Nrs/Ns

Recall: Ratio of selected relevant items to relevant items, i.e., the probability of a relevant item to be selected. R= Nrs/Nr

F1: Combination metric that equals both the weights of precision and recall. F1=(2*P*R)/(P+R)

The goal of the experiment is to test the performance of our prototype in the generation of accurate and relevant content-based recommendations for the users of the system, exclusively considering the mono-disciplinary search. To do so, we have asked a random sample of twelve researchers in the field of Library and Information Science that develop their activity at the University of Granada to evaluate the results provided by the prototype.

One of the premises of the experiment is that at least one of the topics defined for a relevant resource and one of the experts' preferences must be constraint to the same sub-domain of the thesaurus. In such a way, we can leverage a better terminological control on subjects and preferences and extrapolate the output data to the whole thesaurus. In this case, the sub-domain selected is "*Archival science*", which is composed of 96 different concepts. We also require two more elements:

- an RSS feed containing 30 items extracted from the E-LIS (2009) open access repository, from which only 10 of them are semantically relevant (i.e. with at least one subject pertaining to the selected sub-domain).
- a set of profiles with at least one preference pertaining to the targeted sub-area.

The prototype is set to recommend up to 10 resources and then users are asked to assess the results by explicitly stating which of the recommended items they consider are relevant. The results of the experiment are shown in table 3.

Precision, recall and F1 for each user are shown in table 4 (in percentage) and represented in the graph in figure 4. The average outcomes reveal a quite good performance of the prototype.

CONCLUSION

In this paper we have presented a multi-agent filtering and recommender system prototype for digital libraries designed to be used by the scholarly community, that provides an integrated solution to minimize the problem of access relevant information in vast document repositories and finding new research colleagues.

Table 3. Experimental data.

	User1	User2	User3	User4	User5	User6	User7	User8	User9	User10	User11	User12
Nrs	6	5	3	6	4	5	5	4	6	3	7	6
Nrn	2	3	2	1	2	3	2	2	2	2	1	2
Nis	4	5	7	4	6	5	5	6	2	7	3	4
Nr	8	8	5	7	6	8	7	6	8	5	8	8
Ns	10	10	10	10	10	10	10	10	10	10	10	10

Table 4. Detailed experimental outcomes

%	User1	User2	User3	User4	User5	User6	User7	User8	User9	User10	User11	User12	Aver.
P	60.00	50.00	30.00	60.00	40.00	50.00	50.00	40.00	60.00	30.00	70.00	60.00	50.00
R	75.00	62.50	60.00	85.71	66.67	62.50	71.43	66.67	75.00	60.00	87.50	75.00	70.66
F1	66.67	55.56	40.00	70.59	50.00	55.56	58.82	50.00	66.67	40.00	77.78	66.67	58.19

Figure 4. Precision, recall and F1

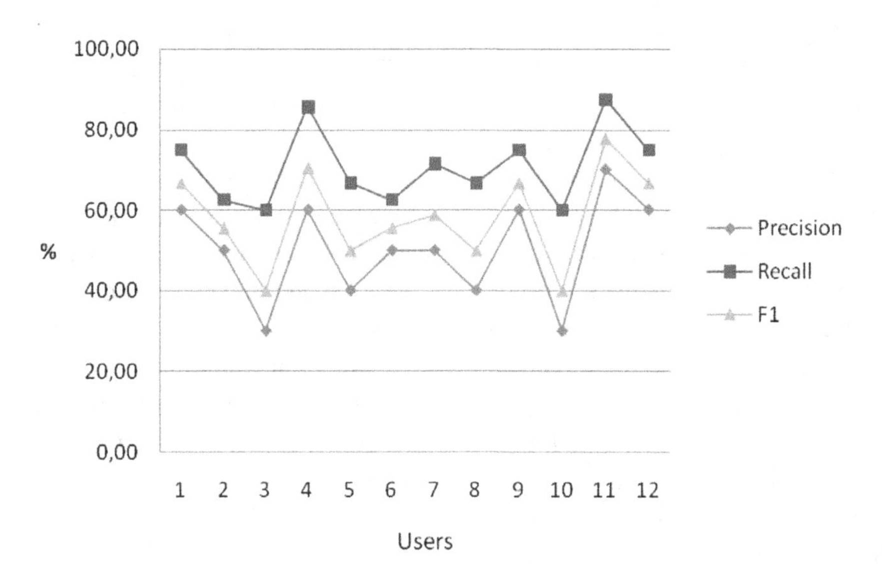

The prototype combines Semantic Web technologies and fuzzy linguistic modeling techniques to improve communication processes and user-system interaction. The use of natural language processing techniques and formal concept analysis facilitates the semiautomatic generation of thesauri from a documental corpus, since they allow detecting descriptors and, in combination with other lexical resources, identifying the semantic relations among them (synonymy, hyponymy and hyperonymy relations in particular). These thesauri can later be used as tools for the semiautomatic indexation of resources and to generate alerts and recommendations.

The system is able to generate both *monodisciplinary* recommendations (to deepen into users' specialization area) and *multidisciplinary* recommendations, which allow users eliciting resources whose topics are tangentially related to their preferences.

The prototype makes possible for researchers to uncover implicit social networks, which relate them with other researchers from different domains, thus easing the task of forming multidisciplinary working groups. Nevertheless, this implies that the system should apply privacy policies to protect those individuals that prefer to be *invisible* for the rest of users.

The system has been evaluated and experimental results show that the model is reasonably effective in terms of precision and recall, although further detailed evaluations may be necessary.

ACKNOWLEDGMENT

This work has been supported by FEDER funds in the National Spanish Projects TIN2007-61079, PET2007_0460 and FOMENTO-90/07.

REFERENCES

Angehrn, A. A., Maxwell, K., Luccini, A. M., & Rajola, F. (2008). Designing collaborative learning and innovation systems for education professionals. In Miltiadis, D., Lytras, J. M., Carroll, E. D., & Tennyson, R. D. (Eds.), *Emerging technologies and Information Systems for the knowledge society. (LNCS 5288)* (pp. 167–176). Heidelberg, Germany: Springer. doi:10.1007/978-3-540-87781-3_19

Aussenac-Gilles, N., Biébow, B., & Szulman, S. (2000). Revisiting ontology design: A method based on corpus analysis. *Proceedings of the 12th International Conference on Knowledge Engineering and Knowledge Management, methods, models and tools* (LNAI 1937), (pp. 172-188). Heldelberg, Germany: Springer-Verlag.

Beckett, D. (Ed.). (2004). *RDF/XML syntax specification*. Retrieved May 12, 2010, from http://www.w3.org/TR/rdf-syntax-grammar/

Beged-Dov, G., Bricley, D., Dornfest, R., Davis, I., Dodds, L., & Eisenzopf, J. ... E. van der Vlist (Eds.). (2001). *RDF site summary (RSS) 1.0*. Retrieved May 15, 2010, from http://web.resource.org/rss/1.0/spec

Berners-Lee, T., Hendler, J., & Lassila, O. (2001, May). The Semantic Web: A new form of Web content that is meaningful to computers will unleash a revolution of new possibilities. *The Scientific American*. Retrieved May 10, 2010, from http://www.sciam.com/article.cfm?id=the-semantic-web

Borgman, C. L. (2007). *Scholarship in the digital age: Information, infrastructure, and the Internet*. Cambridge, MA: MIT Press.

Cao, Y., & Li, Y. (2007). An intelligent fuzzy-based recommendation system for consumer electronic products. [Amsterdam, The Netherlands: Elsevier.]. *Expert Systems with Applications*, *33*(1), 230–240. doi:10.1016/j.eswa.2006.04.012

Cleverdon, C. W., Mills, J., & Keen, E. M. (1966). *Factors determining the performance of indexing systems*, vol. 2, test results. Cranfield: ASLIB Cranfield Project.

E-LIS. (2009). *Home page*. Retrieved May 10, 2010, from http://eprints.rclis.org/

Gruber, T. R. (1995). Toward principles for the design of ontologies used for knowledge sharing. *International Journal of Human-Computer Studies*, *43*(5-6), 907–928. doi:10.1006/ijhc.1995.1081

Guarino, N. (1998). Formal ontology and Information Systems. In Guarino, N. (Ed.), *Formal ontology in Information Systems* (pp. 3–17). Amsterdam, The Netherlands: IOS Press.

Hearst, M. A. (1992). Automatic acquisition of hyponyms from large text corpora. *Proceedings of the 14th conference on Computational Linguistics*, New Jersey, USA, (pp. 539-545).

Hendler, J. (2001). Agents and the Semantic Web. *IEEE Intelligent Systems*, (March-April): 30–37. doi:10.1109/5254.920597

Herrera, F., Herrera-Viedma, E., & Verdegay, J. L. (1996). Direct approach processes in group decision making using linguistic OWA operators. *Fuzzy Sets and Systems*, *79*(2), 175–190. doi:10.1016/0165-0114(95)00162-X

Herrera, F., & Martinez, L. (2000). A 2-tuple fuzzy linguistic representation model for computing with words. *IEEE Transactions on Fuzzy Systems*, *8*(6), 746–752. doi:10.1109/91.890332

Herrera-Viedma, E., Peis, E., Morales-del-Castillo, J. M., & Anaya, K. (2007). Improvement of Web-based service Information Systems using fuzzy linguistic techniques and Semantic Web technologies. In Liu, J., Ruan, D., & Zhang, G. (Eds.), *E-service intelligence: Methodologies, technologies and applications* (pp. 647–666). Heidelberg, Germany: Springer-Verlag.

Isaac, A., & Summers, E. (2008). *SKOS Simple Knowledge Organization System primer: W3C working draft 29 August*. Retrieved May 22, 2010, from http://www.w3.org/TR/skos-primer/

Khan, L., & Luo, F. (2002). Ontology construction for information selection. *Proceedings of the 14th IEEE International Conference on Tools with Artificial Intelligence*, (pp. 122-131). New York, NY: IEEE Computer Society.

McGuinness, D. L., & van Harmelen, F. (Eds.). (2004). *OWL Web ontology language overview*. Retrieved May 17, 2010, from http://www.w3.org/TR/2004/REC-owl-features-20040210/

Miller, G. A. (1995). WordNet: A lexical database for English. *Communications of the ACM*, *38*(11), 39–41. doi:10.1145/219717.219748

Missikof, M., Navigli, R., & Velardi, P. (2002). Integrated approach to Web ontology learning and engineering. *IEEE Computer*, *35*(11), 60–63.

Oldakowsky, R., & Byzer, C. (2005). *SemMF: A framework for calculating semantic similarity of objects represented as RDF graphs*. Retrieved May 23, 2010, from http://www.corporate-semantic-web.de/pub/SemMF_ISWC2005.pdf

Palmer, C. L., Teffeau, L. C., & Pirmann, C. M. (2009). *Scholarly information practices in the online environment: Themes from the literature and implications for library service development.* Report commissioned by OCLC Research. Retrieved May 25, 2010, from http://www.oclc.org/programs/publications/reports/2009-02.pdf

Pedraza-Jiménez, R., Valverde-Albacete, F., & Navia-Vázquez, A. (2006). A generalisation of fuzzy concept lattices for the analysis of Web retrieval tasks. *Proceedings of the 6th International Conference on Information Processing and Management of Uncertainty in Knowledge-Based Systems*, Paris, France. New York, NY: IEEE

Popescul, A., Ungar, L. H., Pennock, D. M., & Lawrence, S. (2001). Probabilistic models for unified-collaborative and content-based recommendation in sparse-data environments. In J. S. Breese & D. Koller (Eds.), *Proceedings of the 17th Conference on Uncertainty in Artificial Intelligence (UAI)*, (pp. 437-44). Massachusetts: Morgan Kaufmann.

Salton, G. (1971). The Smart retrieval system–experiments. In Salton, G. (Ed.), *Automatic document processing*. New Jersey: Prentice–Hall.

Sarwar, B., Karypis, G., Konstan, J., & Riedl, J. (2000). Analysis of recommendation algorithms for e-commerce. In A. Jhingran, J. M. Mason & D. Tygar (Eds.), *Proceedings of ACM E-Commerce 2000 conference*, (pp. 158-167). New York, NY: ACM.

Schein, A. I., Popescul, A., & Ungar, L. H. (2002). Methods and metrics for cold-start recommendations. In K. Jarvelin, M. Beaulieu, R. Baeza-Yates & S. H. Myaeng, (Eds), *Proceedings of the 25'th Annual International ACM SIGIR Conference on Research and Developmentin Information Retrieval (SIGIR 2002)*, (pp. 253-260). New York, NY: ACM.

van Rijsbergen, C. J. (1979). *Information retrieval*. London, UK: Butterworths.

Wordnet. (2009). *Homepage*. Retrieved May 07, 2010, from http://wordnet.princeton.edu/

Yager, R. R. (2007). Centered OWA operators. *Soft Computing*, *11*(7), 632–639. doi:10.1007/s00500-006-0125-z

Zadeh, L. A. (2003). The concept of a linguistic variable and its applications to approximate reasoning. *Information Sciences*, *9*(3), 43–80.

Chapter 7
Domain-Specific Ontologies Trading for Retrieval and Integration of Information in Web-Based Information Systems

José-Andrés Asensio
University of Almería, Spain

Javier Criado
University of Almería, Spain

Luis Iribarne
University of Almería, Spain

Nicolás Padilla
University of Almería, Spain

ABSTRACT

This chapter introduces the use of domain-specific ontologies through Web trading services as a mechanism for retrieval and integration of information between different systems or subsystems. This mechanism is based on a three-level data architecture, which can be demonstrated by the use of trading service. This architecture includes data at its first level, meta-information on its second level in order to facilitate the processes of retrieval of information, and meta-meta-information at its third level to facilitate the integration of information-through-trading-services. This proposal is a new approach to the process of retrieval and integration of information for Web-based Information Systems (WIS). This chapter presents a case study for a WIS application of an Environmental Management Information System (EMIS), called SOLERES.

DOI: 10.4018/978-1-61520-921-7.ch007

INTRODUCTION

With the progress of technology, Information Systems (IS) must be more and more flexible, easily scalable, and providing full accessibility to its users, enabling the collaborative work, and facilitating the access to the information and so on, an information which can come from different sources, to help them in the decision-making process, etc., that is called "convergent systems". All these features can be achieved basically through the use of: (a) a common vocabulary between all systems, provided by the ontologies, and (b) a certain capacity to mediate between systems, which favours communication, negotiation, coordination, etc. In this way, current WIS are developed under open and distributed paradigms, and are based on rules and standards that enable interaction and interconnection in real time (Xiao-Feng *et al.*, 2006). The main tools used for implementing these systems are XML technologies, semantic web, techniques of query and extraction of information, or web data-mining, among others (Taniar *et al.*, 2004).

Usually, web users have a great volume of information and use search engines, websites, digital libraries, and other systems, for querying and retrieval of information. Even so, users have to make an effort to locate the relevant content. In an attempt to solve this problem, a wide variety of techniques have been developed in Web-Based Information Systems (WIS), related to searching, recovering (Carrillo-Ramos *et al.*, 2006) and filtering of information. Trader can be considered as another solution to perform these processes in open and distributed systems (Trader, 1996). Although traditionally used as middleware to provide interoperability among objects, they can be easily adapted to replace these objects for information, as it will be exposed in the development of this work.

The rest of the chapter is structured as follows. Section 2 provides a brief description of the state of the art in this field. Section 3 provides an overview of the architecture of data with the three levels mentioned above, and its justification with the use of the trading service for retrieval and integration of information. Then, in Section 4 this is applied in a sample scenario. And finally, conclusions are presented in Section 5.

RELATED WORK

This section reviews the most important work done in the field of the application of domain-specific ontologies and trading services to the processes of retrieval and integration of information.

Some works suggest the use of ontologies in the process of searching and retrieval of information in order to improve this process (Chien *et al.*, 2007; Chien *et al.*, 2010). For example, in (Qi *et al.*, 2008) they use a domain-specific ontology for the query of web documents. There are also works that propose ontologies as a mechanism for the integration between different systems. While in (Dartigues *et al.*, 2007) an approach to the integration between a Computer-Aided Design-(CAD) system and a Computer-Aided Process Planning (CAPP) system is described, in (Rajapakse *et al.*, 2008) they propose another domain-specific ontology for the integration of information-databases with scientific literature and the subsequent search. For its part, (Sousan *et al.*, 2007) provides an Intelligent-Web-Service based on semantics for the retrieval and integration of information that uses a domain-specific ontology, which contains the user preferences and is developed and refined by the searching that they perform.

As far as trading services are concerned, there are works that use a domain ontology to define the vocabulary that describes the services, and a trader that facilitates the finding of service which best fits the preferences of the user (Sora *et al.*, 2009). In (Zein *et al.*, 2006) they go a step further and propose an ontology-based-trader to

Figure 1. Double dimension of the common use of trading services and ontologies: (a) information retrieval, and (b) information integration

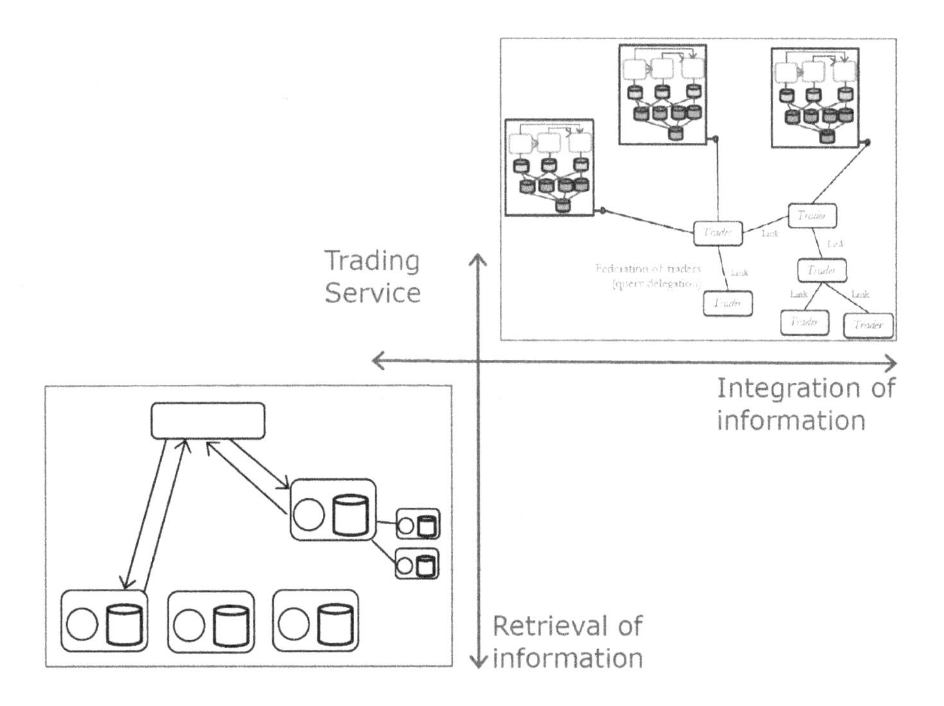

describe both static and dynamic properties and service interfaces.

Although there are works developed in this area, it seems that there are no studies that combine the use of a trading service and domain-specific ontologies in the process of retrieval and integration of information (Figure 1). As it has been seen, on the one hand, domain-specific ontologies are used in searching, retrieval and integration of information and, on the other hand, trading services use ontologies for describing services, but not for describing information. It is at this point where the work presented in this chapter is focused.

MULTI-LEVEL KNOWLEDGE REPRESENTATION AND TRADING

As it has been mentioned up to this moment, the purpose of this chapter is to propose a new mechanism for retrieval and integration of information on the web, either among different systems or between the different subsystems that make up a system. This mechanism is based on the use of domain-specific ontologies and trading service. This involves the use of a data architecture with three levels, which are described and demonstrated using Figure 2.

As is shown in Figure 2, the existence of one or more systems or subsystems (for simplicity, henceforth all will be considered as systems) that use a set of data is assumed. These data can be part of the system, or can come from external data sources, and make up the first level of the data architecture. To speed up the process of retrieval information it is not necessary to work directly with the data, because of the volume of data that it can produce; so a second level appears and has been called meta-information. This meta-information has a smaller volume and, although many times it is necessary to go to the level of data to solve a query, many other queries are solved at

Figure 2. Data architecture with three levels

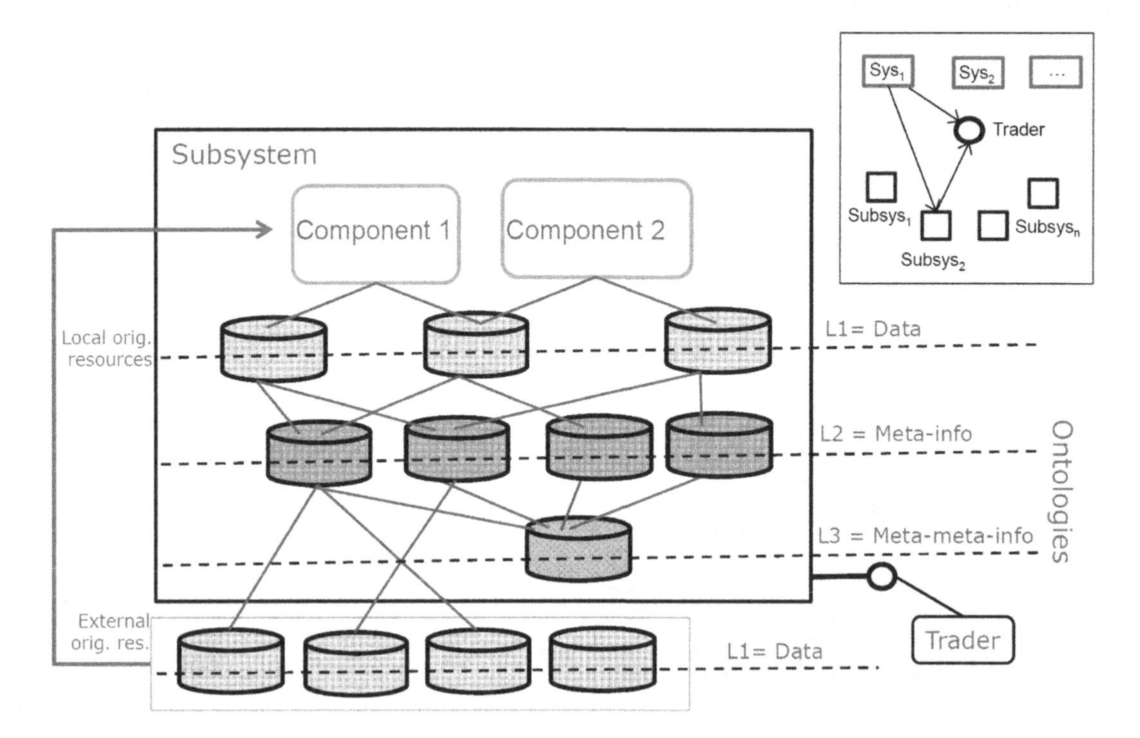

this level. However, if it is necessary to go down to the level of data, the level of meta-information acts as a filter, reducing the volume, and consequently, reducing the execution time of the query. If there were only one system, with these two levels, it would be more than enough to the searching and retrieval of information; but the purpose of this work is also to facilitate the possibility of carrying out the integration between two or more systems. For this reason, a new level is included in the architecture, and has been called level of meta-meta-information. Some information that relates data from different systems can be picked up with it. This level of information is used by the trading service.

The trading service that has been proposed is based on the ODP trading-service-specification but adapted to the management of information templates: meta-meta-information templates, rather than service templates. Specifically, the responsible web component for providing this service, which is called trader, would have all the meta-meta-information templates from the systems associated with it. Federation could even be carried out between these types of components, but it is not within the scope of this work. Figure 3 shows a generic scenario in which the operation of the trading service can be seen clearly.

In Figure 3 there are, roughly, a query manager, a trader (in order to simplify the scenario, it is going to be reduced to the use of a single trader, regardless of the federation) and different systems. Each system has a data repository (either internal or external), and also two internal repositories, one of meta-information over data, and another of meta-meta-information, with information that relates its meta-information to other systems' meta-information.

In short, it is integrating data from different systems. For its part, the trader has its own meta-meta-information repository that includes all associated systems' templates. As can be seen,

Figure 3. Operation of the trading service

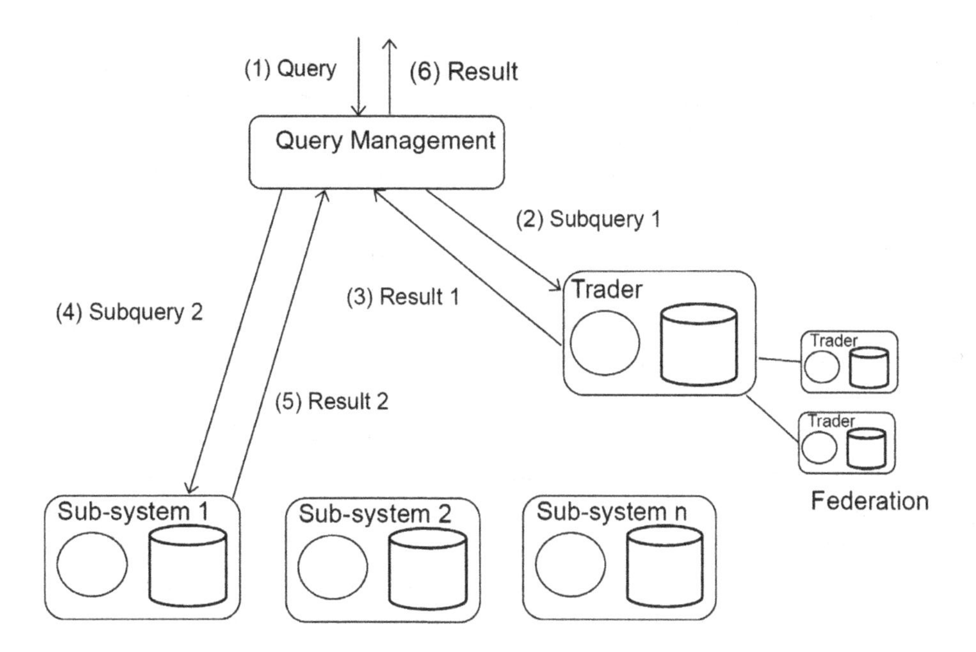

this meta-meta-information is duplicated, and the reason for this is to have redundancy to any failure that the trader may present. And finally, the query manager is responsible for receiving queries from users, solving them with the help of the trader, and the systems described above, and returning the results to the users.

Once the query manager receives a query from a user, three main situations can arise. (a) If the query can be resolved directly with the trader's meta-meta-information templates, the manager sends the query to the trader, which runs the query against its repository, returns the results to the manager, and the manager treats them, returning the final result to the user. (b) If the query can be solved with the system's or systems' meta-information templates, the manager sends them the query, they make the necessary queries over their meta-information repositories and then return their results; the manager treats them, again returning a final result to the user. (c) If the query is more complex and the systems' meta-meta-information and meta-information must both be solved, firstly, the query manager decomposes the query into two subqueries. Next, it sends the first subquery to the trader, which runs the pertinent query through its repository and returns its results to the manager. With these partial results, the manager builds the second subquery and sends it to one or more systems, which make the necessary queries over their meta-information repositories, and returns its results back to the manager. Finally, it treats them and returns a final result to the user.

CASE STUDY: SOLERES SYSTEM

As it has been mentioned above, EMIS are classified as WIS. These systems are a special type of Geographic Information Systems (GIS), where the granularity of the environmental information is much higher. An example of EMIS is the SOLERES system, a spatio-temporal environmental management system based on neural networks, agents and software components (ACG/CEG, 2008; Asensio *et al.*, 2008; Asensio *et al.*, in press; Iribarne *et al.*, 2008; Padilla *et al.*, 2008), which will be used as a case study. The main purpose of this system is the study of a framework for the integration of all the above disciplines, using the environment as application domain. In this section everything mentioned in the previous section will be applied within the SOLERES system, that is, a data architecture with three levels common for different subsystems will be defined and it will be demonstrated how a trader allows the integration of these subsystems and supports the process of searching and retrieval of the environmental information, in this case ecological classifications and satellite image classifications.

Starting with the data architecture, each subsystem has some data sources, initially external to the system, which collect all data records from ecological classifications and satellite image classifications. This represents a great volume of information. These data sources integrate the first level of the architecture. But apart from these data sources, each subsystem manages two local repositories of environmental information. The first one contains meta-information from the domain data sources and is called Environmental Information Map (EIM) or EIM template. EIM templates represent the second level of data within the architecture. The second repository contains meta-meta-information and is called Environmental Information metaData (EID) or EID template. For each EIM template there is an EID template. These templates represent the third level of data, and contain the most representative meta-information of the EIM templates used in the process of searching and retrieval of information, as well as another set of information necessary for the integration of different subsystems. Both templates, EIM and EID templates, have been designed as ontologies and expressed in Ontology Web Language (OWL) (Asensio *et al.*, in press; Padilla *et al.*, 2008).

As mentioned previously, EIM templates represent the meta-information related to the ecological classifications and satellite image classifications. Figure 4 shows partially the diagram of this template. Here you can see how ecological classifications (Ecological_classification) are made from cartographic maps (Cartography), while satellite image classifications (Image_classification) obviously require such images (Satellite_image). On the one hand, cartographic maps (Cartography) include both geographical information (Geography), necessary to establish the georeferencing properties of the map (coordinates, area, perimeter...), and ecological maps (Layer). These consist of ecological variables (Variable), for example, temperature, precipitation and so forth, and are stored in files (Resource) of different types. On the other hand, images (Satellite_image) are taken by satellites (Satellite), which use sensors or instruments (Instrument) to obtain them with a certain resolution (Resolution) and using a set of bands (Band). The resolution contains spectral, radiometric, temporal and spatial information. The satellite images are also stored in files.

The aim of both ecological classification and satellite image classification, consists of dividing the map or image into a discrete set of classes with similar characteristics. Both classifications are made by one or more technicians (Technician) at a certain time (Time). But while ecological classification is based on the generation of grids (Grid) and on a hierarchy (Hierarchy), in which new maps and new variables can be generated from some maps and variables, the satellite image classification uses a training set (Training_set) and a method of performance evaluation (Performance_indicator) to get classes (Classes), which generate the classified images (Classified_image), finally stored in files (Resource).

In the diagram of the EID template, shown in Figure 5, it can be observed that the most important information from the classifications, both ecological and satellite images, stored in the EIM templates, appears joined in a single element (Classification): geography (Geography), technicians involved (Technician), realization time (Time) and ecological maps (Layer) with their respective variables (Variable) or satellite images (Satellite_image) containing the information of

Figure 4. Diagram of the EIM template

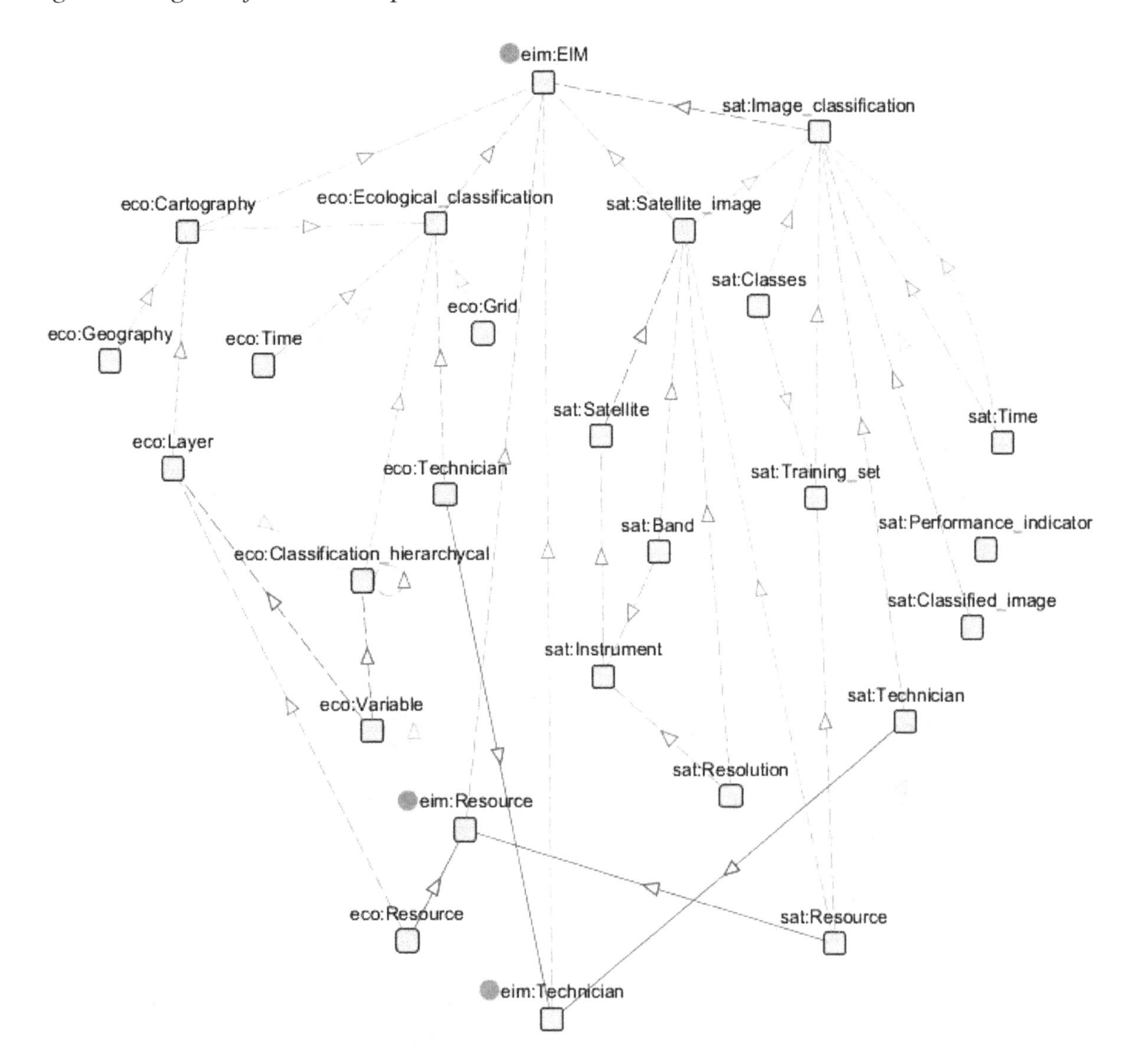

the bands (Band). All these elements collect the minimum important information from the EIM template and add some attributes used by the trader in the process of searching and retrieval of information, as it will be seen now.

Figures 6 and 7 show, respectively, the contents of an EIM template and its corresponding EID template.

Then the process for searching and recovering will be analyzed. As has been explained in the previous section, a new component appears in the system, the trader, which will have a direct and positive influence in the query resolution. At first, all subsystems are associated with the trader and replicate their EID templates, generated from each EIM template, stored in its local repository. Then, when the manager receives a query, such as "what is the range of values for the climate variable of average annual rainfall (low) from maps used in ecological classifications made in the province of Almeria between 2005 and 2008, where regions are classified as dry?", the following process oc-

Figure 5. Diagram of the EID template

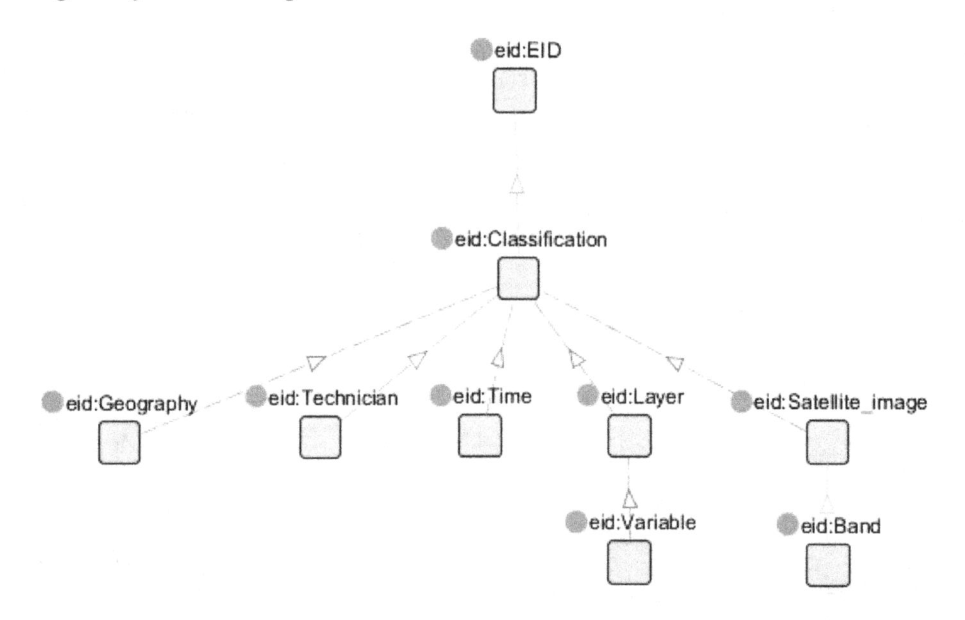

curs: (a) the manager divides the query into two subqueries. The first of them is "what are the EIM templates that refer to ecological classifications (Classification) performed in the province of Almería (Geography) from 2005 (Time) and 2008 (Time), where regions are classified as dry (Variable)?" and it would be sent to the trader. (b) The trader runs this query over its repository and tells the query manager those identifiers of the EIM templates that meet the specified criteria. (c) The manager generates a second query, "what is the range of values for the climate variable of average annual rainfall (low) (Variable) from maps (Layer) contained in the templates with those identifiers...?", only addressed to those subsystems that have the searched templates. (d) These subsystems perform the concrete query over their EIM templates and return their results to the manager. (e) Finally, the manager treats all the results obtained and returns a final result to the user.

If the trader does not take part in this process, each and every one of the subsystems would have to be inspected to search the requested information by the user. Here the improvement can be seen clearly.

CONCLUSION

Due to the growth that the web is having in recent years as well as the increase number of Information Systems and the volume of information they handle, even more efficient mechanisms for information-retrieval and integration of information are necessary. For this reason, the purpose of this work has been to provide a new approach based on the use of a data architecture with three levels, based on domain-specific ontology and web trading service as mechanism for retrieval and integration of information. This architecture presents data at its first level, meta-information to facilitate the processes of searching and retrieval of information at the second level, and meta-meta-information to facilitate the integration process through the trading service at its third level.

This approach has been successfully applied in an Environmental Management Information

Figure 6. Instance of an EIM template

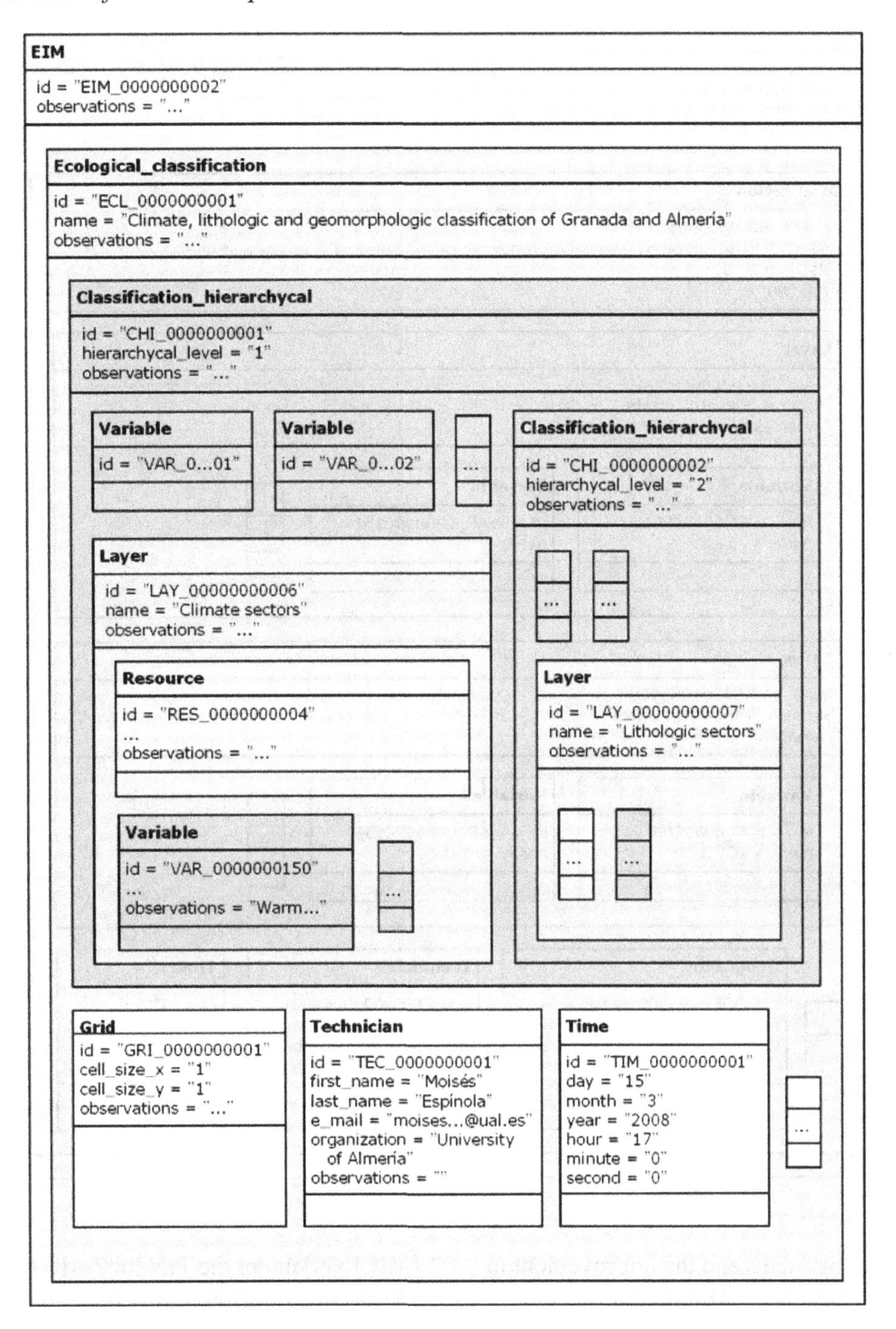

System (EMIS), called SOLERES, a spatio-temporal environmental management system based on neural networks, agents and software components. The system, composed by different subsystems, has been used as the base to extend the previous architecture with a trading service inspired in ODP and verifying both, the integra-

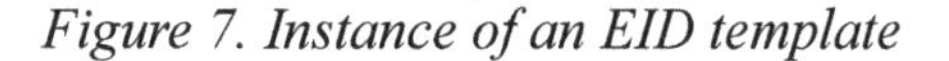

Figure 7. Instance of an EID template

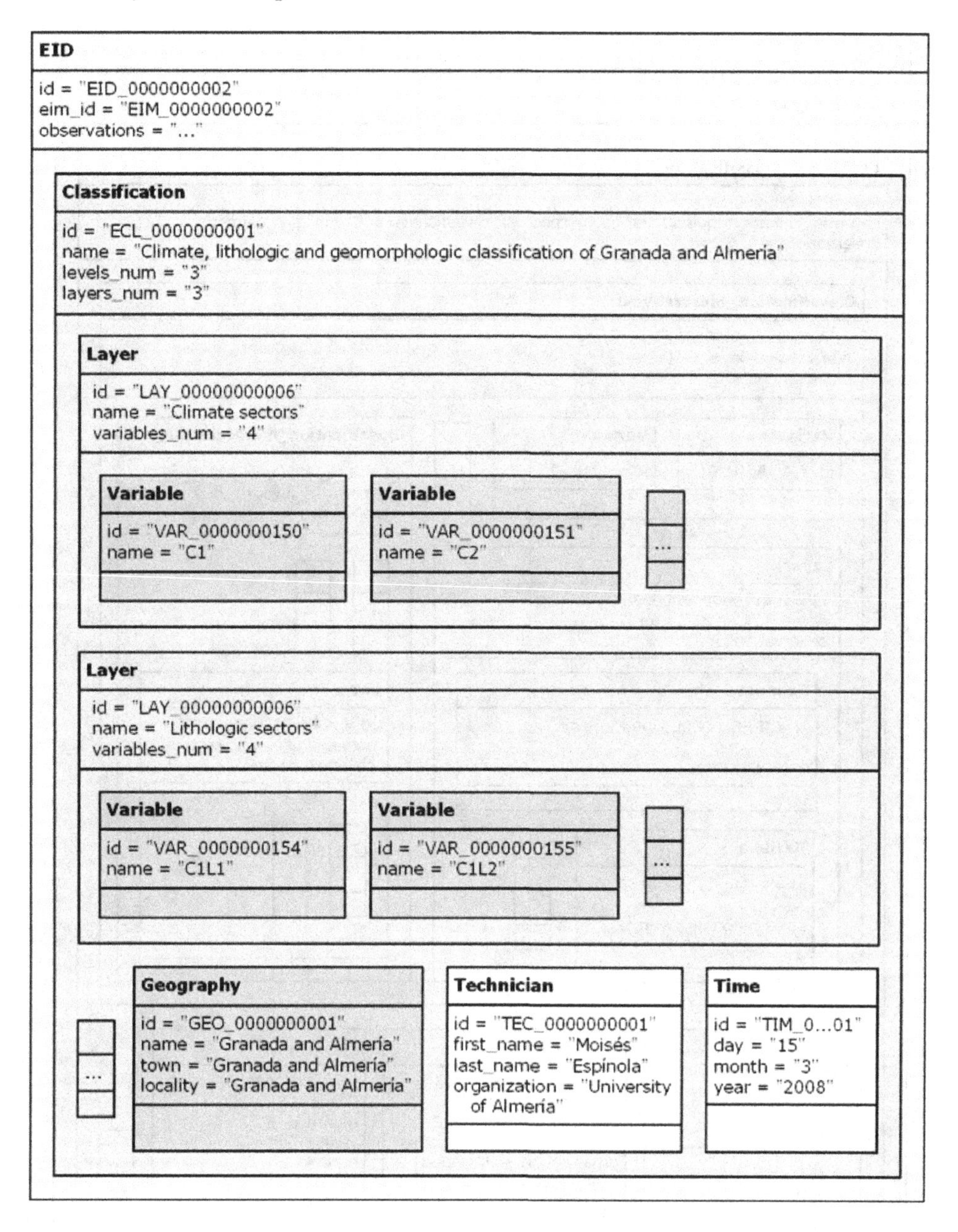

tion of these subsystems and the improvement in the process of searching information.

ACKNOWLEDGMENT

This work was funded by the EU (ERDF) and the Spanish Ministry of Science and Innovation (MICINN) under the TIN2007-61497, TIN2010-15588, and TRA2009-0309 Projects, http://www.ual.es/acg/soleres.

REFERENCES

ACG/CEG. (2008). *The SOLERES R&D Project: A spatio-temporal environmental management Information System based on neural-networks, agents and software components*. (Technical report, Applied Computing Group (ACG) and Computers and Environmental Group (CEG), University of Almeria, Spain). Retrieved from http://www.ual.es/acg/soleres

Asensio, J. A., Iribarne, L., Padilla, N., & Ayala, R. (2008). Implementing trading agents for adaptable and evolutive COTS components architectures. In *Proceedings of the International Conference on e-Business* (pp. 259-262). Porto, Portugal.

Asensio, J. A., Padilla, N., Iribarne, L., Muñoz, F., Ayala, R., Cruz, M., & Menenti, M. (in press). A MDE-based satellite ontology for environmental management systems. *Information Technology. Information Systems and Knowledge Management*.

Carrillo-Ramos, A., Gensel, J., Villanova-Oliver, M., & Martin, H. (2006). *Adapted information retrieval in Web Information Systems using PUMAS*. (LNCS 3529), (p. 243).

Chien, B.-C., Hu, C.-H., & Ju, M.-Y. (2007). Intelligent information retrieval applying automatic constructed fuzzy ontology. In *proceedings of 6th International Conference on Machine Learning and Cybernetics*, 1-7 (pp. 2239-2244). Hong Kong.

Chien, B.-C., Hu, C.-H., & Ju, M.-Y. (2010). Ontology-based information retrieval using fuzzy concept documentation. *Cybernetics and Systems*, *41*(1), 4–16. doi:10.1080/01969720903408565

Dartigues, C., Ghodous, P., Gruninger, M., Pallez, D., & Sriram, R. (2007). CAD/CAPP integration using feature ontology. *Concurrent Engineering-Research and Applications*, *15*(2), 237–249. doi:10.1177/1063293X07079312

Iribarne, L., Asensio, J. A., Padilla, N., & Ayala, R. (2008). Modelling a human-computer interaction framework for Open EMS. The Open Knowledge Society: A Computer Science and Information Systems Manifesto. In *proceedings of the 1st World Summit on the Knowledge Society* (pp. 320-327). Athens, Greece.

Padilla, N., Iribarne, L., Asensio, J. A., Muñoz, F., & Ayala, R. (2008). Modelling an environmental knowledge-representation system. Emerging Technologies and Information Systems for the Knowledge Society. In *proceedings of the 1st World Summit on the Knowledge Society* (pp. 70-78). Athens, Greece.

Qi, Z., Zheng, Y., & Jiang, X. (2008). An approach for domain ontology-based semantic retrieval. In *proceedings of 3rd China-Ireland International Conference on Information and Communications Technologies* (pp. 185-189). Beijing.

Rajapakse, M., Kanagasabai, R., Ang, W. T., Veeramani, A., Schreiber, M. J., & Baker, C. J. O. (2008). Ontology-centric integration and navigation of the Dengue literature. *Journal of Biomedical Informatics*, *41*(5), 806–815. doi:10.1016/j.jbi.2008.04.004

Sora, I., Todinca, D., & Avram, C. (2009). *Translating user preferences into fuzzy rules for the automatic selection of services*. In the 5th International Symposium on Applied Computational Intelligence and Informatics, IEEE. (pp. 487-492). Timisoara, Romania.

Sousan, W. L., Payne, M., Nickell, R., & Zhu, Q. (2007). *Metadata (ontology) incremental building and refinement agents*. In 2007 International Conference on Integration of Knowledge Intensive Multi-Agent Systems, IEEE. (pp. 127-132). Waltham.

Taniar, D., & Rahayu, J. (2004). *Web Information Systems*. Hershey, PA: IGI Global.

Trader, I. (1996). *ISO/IEC DIS 13235-1: IT-ODP trading function-part 1: Specification.*

Xiao-Feng, M., Bao-Wen, X., Qing, L., Ge, Y., Jun-Yi, S., Zheng-Ding, L., & Yan-Xiang, H. (2006). A survey of Web Information Technology and application. [Wuhan University.]. *Journal of Natural Sciences*, *11*(1), 1–5.

Zein, O. K., Kermarrec, Y., & Salaun, M. (2006). An approach for discovering and indexing services for self-management in autonomic computing systems. *Annales des Télécommunications*, *61*(9-10), 1046–1065.

Chapter 8
Technology Engineering for NPD Acceleration:
Evidences from the Product Design

Nouha Taifi
University of Salento, Italy

Eliana Campi
University of Salento, Italy

Valerio Cisternino
University of Salento, Italy

Antonio Zilli
University of Salento, Italy

Angelo Corallo
University of Salento, Italy

Giuseppina Passiante
University of Salento, Italy

ABSTRACT

In complex environments, firms adopt continuously new IT-based systems and tools for knowledge management, otherwise knowledge can be dispersed or lost. And as a part of the new product development process, the product design is one of the most crucial phases for the relevance of its data and information and for the importance of the new knowledge creation of its designers and engineers. This chapter argues, through a conceptual model, the strategic role of the integration of knowledge management systems and special communities for the acceleration of the new product development process and presents an ontology-based knowledge management system and its application in the context of a community of automotive designers. More precisely, the issue management, based on this engineered IT-system, will accelerate and optimize the product design phase and knowledge sharing among the designers and engineers.

DOI: 10.4018/978-1-61520-921-7.ch008

INTRODUCTION

In the actual environment, new product development has emerged as the main objective of many businesses and organizations. It is a process characterized by complex steps that need to be organized, managed and supported by knowledge management. Knowledge is the key success factor for the achievement of high economic performance for organizations and knowledge management leads to an efficient knowledge creation, reuse and sharing. Managing knowledge for new product development is a complex process that incite organizations to find new methods to optimize it. The organizations' objectives is to coordinate the management of knowledge through optimal solutions as the creation of special groups, entities, units and teams, and to support them through the use of information systems based on technologies.

Merging organizational structures, knowledge management and technological systems is the biggest challenge of most organizations. And the reason for that is the complexity of the new product development process and the willingness to accelerate it. Our goal is to bring together the three areas of research, to connect them with NPD and to end up with a proposed solution. In what follows, we will talk about the communities of practice (CoP) as strategic entities for knowledge creation and sharing and the knowledge management systems (KMS) as their enablers of knowledge storage and reuse. We then propose a framework for the connection of the CoP and KMS and the extent to which it contributes to the acceleration of the NPD process. Next, we focus on an adequate IT-based method to achieve NPD acceleration that is based on semantic web and present a case of the management of knowledge related to new complex products design where the method is contextualized and customized.

ORGANIZATIONAL STRUCTURE

The organizational structure plays a strategic role in the context of new product development where the rules and procedures are coordinated among the people involved in the development of a product. Wheelwright and Clark (1992) state that the functional organizational structure is needed for the development of complex products that require specialized knowledge. The functional structure gathers people holding similar positions in an organization, performing the same type of activities, and exploiting the same kind of expertise. Thus, this type of structure facilitates communication and knowledge sharing, allows fast decision making since the involved people share the same interests, and leads to skills development.

In the functional organizational structure, centralization, formalization and integration are important factors influencing it (Germain (1996) and Sciulli (1998)). Nonaka and Takeuchi (1995) indicate that a combination of formal, non-hierarchical and self-organizing structure would improve knowledge creation and sharing capabilities. Besides, integration allows individuals to work interactively and share knowledge. Therefore, decreased centralization (Nonaka and Takeuchi (1995); Hopper (1990)), low formalization (Sivadas & Dwyer, 2000) and integration lead to knowledge sharing and creation for the new product development. The communities of practice (CoP), as organizational structures, are characterized by these factors and enhance knowledge sharing and creation since CoPs are decentralized, mostly informal and used to share knowledge among different individuals sharing the same interests. More precisely, the strategic communities are created for the purpose of the acceleration of new product development (Corallo et al., 2008; Taifi et al., 2008).

KNOWLEDGE MANAGEMENT SYSTEMS

Organizations and firms are urged to accelerate the new product development (NPD) process that is the enabler to their success on the markets (Cooper, 1994). For that, the strategic element is knowledge (Grant, 1996) that is the basis for the exploitation and exploration (March, 1991) of the firms' resources for value creation and high economic performance (Nonaka, 1991; Davenport & Grover, 2001). Thus, organizational knowledge needs to be managed and leveraged. Knowledge management is critical to the firms' success (Wiig, 1995) and in the actual digital era, new forms of systems for knowledge management are born focusing on the creation, gathering, organization and dissemination of the organizational knowledge (Alavi and Leidner, 2001).

The knowledge management systems (KMS) facilitate the integration of dispersed knowledge (Grant, 1996). They are information systems (Benbya, 2008) supporting and focusing on the management of knowledge for the development of collaboration and sharing of good practice. They are used in contexts where specific knowledge needs to be well organized in order to avoid redundancy and save time of exploitation. The KMS based on information technology (IT) are the adequate information systems for knowledge sharing of complex products (Corallo et al., 2009; 2010). The information technologies provide economic value to the organizations by lowering the costs and contributing to the uniqueness of the products. The IT-based KMS are used in various areas of expertise in the new product development process as in the design and engineering.

CONCEPTUAL FRAMEWORK

Connecting KMSand COP for NPD Acceleration

The communities of practice, as an optimal organizational structure, and knowledge management systems, as the right IT-based information systems are two important factors for the acceleration of the new product development process. In order to combine the CoP with the KMS, the introduction of a common language is necessary. The ontology is a language that is the basis of semantic web. It converts human logic and thinking into computer codified language and leads to knowledge codification. According to Bernes-Lee (2000), it constitutes the backbone of the emerging scenario of semantic web, based on the explicit description of web resources' content. Semantic web leads to fast interactions among users willing to share and create new knowledge (Zilli et al, 2008). Ontology is a meta-level description of knowledge representation; it provides search engines with the functionality of a semantic match; it is used to classify knowledge (Guarino, 1997). Ontology can be used by the communities' members to communicate, interact and exchange knowledge (Corallo et al., 2005). It enhances knowledge flows among CoP members and improves the usability of the technology platforms (Cisternino et al., 2008). Therefore, the combination of CoP and KMS through the use of ontology based on the semantic web can lead to knowledge creation and sharing and it can contribute to the acceleration of the NPD process (Figure 1).

Through the creation of ontology-based KMS for CoPs, it is possible to manage and integrate the knowledge of the members in the NPD process. Füller et al. (2006) state that there are many opportunities in the NPD process to integrate the knowledge of IT-based communities. The first phases of the NPD process are the most critical ones in which ideas are generated, new concepts are created then designed and engineered. As

Figure 1. Combining CoP and KMS through ontologies for NPD acceleration

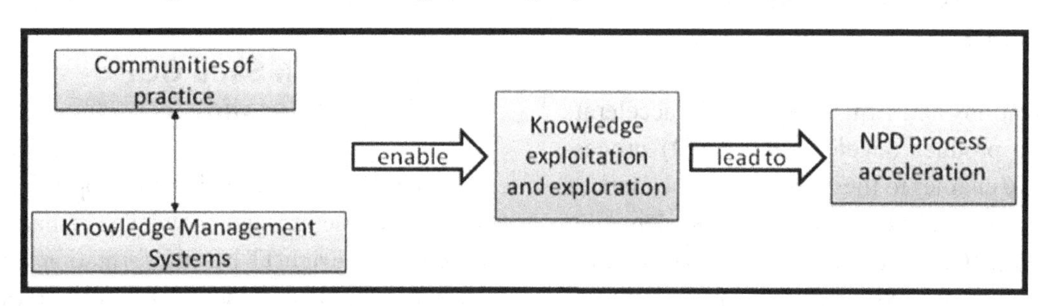

sources of innovation, among others, designers within research and development centers can be critical for the creation of new knowledge and at the same time evaluate it. Therefore, as members of a community, they can contribute with their creativity and problem solving skills for the acceleration of the NPD process (Füller et al., 2006).

ONTOLOGY MODELING BASED ON THE IDEF0 METHODOLOGY

The modeling methodology used is the IDEF0 (Mayer et al., 1992). It is a structured methodology that captures the structure of a system based on the functional specifications in order to understand the way it works (Colquhoun et al., 1993). It is a language based on graphics with a formalism explaining the activity performed, indicating the structural relationships and processing requirements of the components' of a system and including the following: Input (s), Control (s), Output (s) and Mechanism (s).

- The input: is the information or the objects that are sent to the activity.
- The output: is the information or the objects released by the activity.
- The activity: can also be controls or mechanisms for other activities.
- The mechanisms: are the means and resources used for the transformation of the inputs received into outputs (machines, environment, tools, people, etc.)
- The controls: are all the entities that lead the execution of the activity; the controls determine the rules and the conditions required in order to obtain the adequate output.

Carcagni et al. (2008) have used the IDEF0 methodology to create a process for ontology development. The macro-process based on the IDEF0 methodology is designed and named 'Methodology for ontology development'. Through this methodology, all necessary information for the development of the ontology is put in place in order to reach the optimal results.

Besides, the IDEF0 methodology is applied through a top-down and hierarchical breakdown. Its model is therefore composed of several diagrams which are read from left to right and from top to bottom. In these modeling processes, the following is elaborated (Carcagni et al., 2008):

- Analysis of the feasibility: to study the field of interest for which the ontology will be developed, to determine the requirements for the ontology and to benchmark and choose the adequate one.
- Knowledge base representation: which lead to the creation of a precise knowledge base about the field of interest and according to the previously determined requirements, some characteristics for the ontol-

Figure 2. The Technological Architecture of the KMS

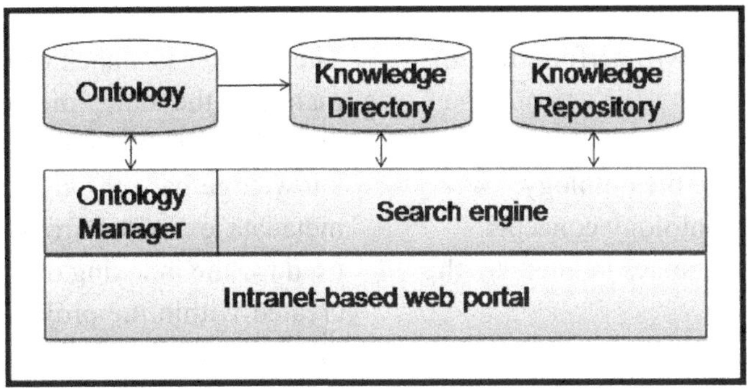

ogy are created, and the parameters for the quantitative and qualitative evaluation are defined.

- Logic modeling: the output of the knowledge base representation is used as an input for the logic modeling in order to obtain a structured and logical knowledge base. Clear semantic relations are defined in order to obtain a logical ontology. The actors involved in this phase are the knowledge capturer, the domain expert and the domain analyst. The logic modeling process consists of the creation of the vocabulary, the thesaurus and the logical ontology.
- Implementation: the formal language expressed in the previous step is translated into machine-readable language, through the use of developed tools and languages supporting ontology development.
- Test: in this step of the ontology modeling, the alignment of the ontology created with the initial requirements is checked. The competency question checking is used to acquire feedback of the metrics for the ontology.
- Development and maintenance: in this step, all the previous steps are analyzed in order to determine the ones in which some development or maintenance need to be made. This step is connected with all the other ones since its output is always an input for each of the other steps for achieving development and maintenance.

ONTOLOGY-BASED KNOWLEDGE MANAGEMENT SYSTEM

The knowledge management system presented below is mainly based on the ontology to categorize, organize and store information. Its technological architecture (Figure 2) is composed of:

- The interface layer; including the Ontology Manager, the Indexer and the Search Engine;
- The data layer; storing data and information;
- The business logic layer; including the softwares supporting the KMS.

Ontology Manager

The Ontology Manager is a tool dedicated to the creation and management of the ontology, on which the semantic indexing will be based. The available tool is OntoMaker, developed by the university research center within the KIWI project (Damiani et al., 2008) related to the creation of a knowledge management system and uses the RDF and OWL schema. The KIWI (Knowledge-based Innovation for the Web Infrastructure) project has contributed to the creation of a new system of technological tools for customized computing based on metadata and ontology.

The Ontology Manager will perform the following actions:

- To create and manage the ontology;
- To modify and add ontology concepts.
- To maintain the instances related to the concepts.
- To manage the relations among the concepts and to create new ones.

These semantic statements in the knowledge base will automatically be updated as the Ontology Manager performs the above functionalities.

The Ontology Manager interface, as represented in the figure 3, can be divided into three sections:

- The up-left section, includes the taxonomy of the ontology elements where it will be possible to add or modify the ontology concepts.
- The up-right section, includes the definition of the selected ontology element in the up-left section.
- The down-right section, includes the list of slots that can be modified for each selected ontology element.

Semantic Indexer

The Indexer is a back office tool that stores and manages the data, metadata and ontology. It links the semantic content to a document in the knowledge base through the association among metadata extracted from an ontology. The tool used for the indexing of the documents has been created within the project KIWI and adapted to this knowledge management system.

The users of the indexer, through a web interface, are the knowledge exporter and the knowledge worker. Respectively, the Administrator is the user that can create and update the ontology and the Power user is the user involved in the feeding of the knowledge base, thus creating a context that supports the processes of knowledge sharing and knowledge transfer among all the users of the system.

The indexer links the knowledge base to the concepts of the domain ontology through the use of simple or complex statements;

- A simple statement is a unique concept, that can be considered as attributions of metadata on the semantic content of the document.

Figure 3. The Ontology Manager Interface

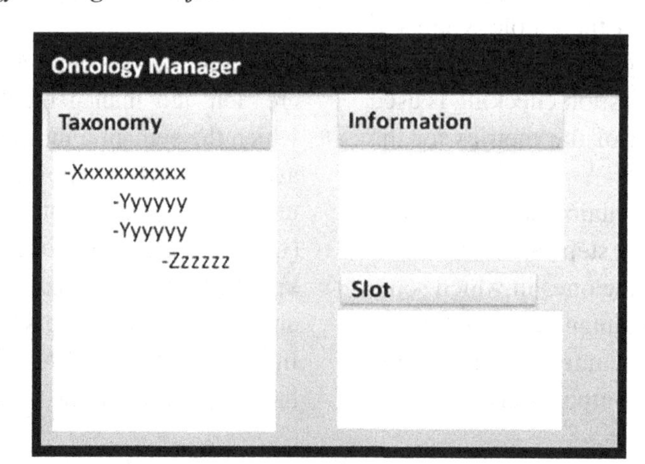

- A complex statement is two concepts emerging from the subject and object of the statement and linked through a relation.

The interface, as represented in the figure 4, can be divided into three parts:

- The up-left section, includes the taxonomy navigation where the taxonomy is uploaded, and explored by the user.
- The up-right section, includes the Knowledge Base Resource Management where it is possible to see the resource that needs to be indexed in XML version and its structure –the resource is subdivided into paragraphs.
- The down-right section, includes the list of assertions about the document currently visualized. In this section, it is also possible to create, modify or cancel an assertion. In order to create it, a new window and the document to be indexed are opened, and the simple or complex assertion is made; This latter will appear in the assertion management section.

Search Engine

The search engine retrieves knowledge from the indexed and stored documents in the knowledge base. The search engine allows two types of navigations:

- Semantic Navigator: To look for and retrieve documents through the searching of ontology' concepts. The documents are extracted through the searching of an ontology concept or a semantic assertion.
- Syntactic Navigator: To retrieve documents according to specific key or synonyms words existing in the text body of the documents.

The semantic navigator, as represented in figure 5, can be divided into four parts:

- Navigation Choice; in which it is possible to choose the type of navigation; semantic in this case;
- Taxonomies Navigation; in which the user can insert, select or navigate the taxonomy;
- KB resource results list; in which the retrieved resources are listed;
- KB resource content visualization; in which it is possible to see the content of the document selected from the resource list.

The syntactic navigator (figure 6) can be divided into four parts:

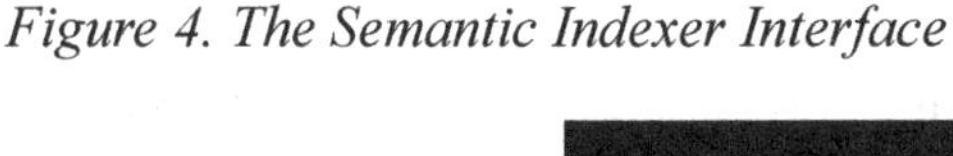

Figure 4. The Semantic Indexer Interface

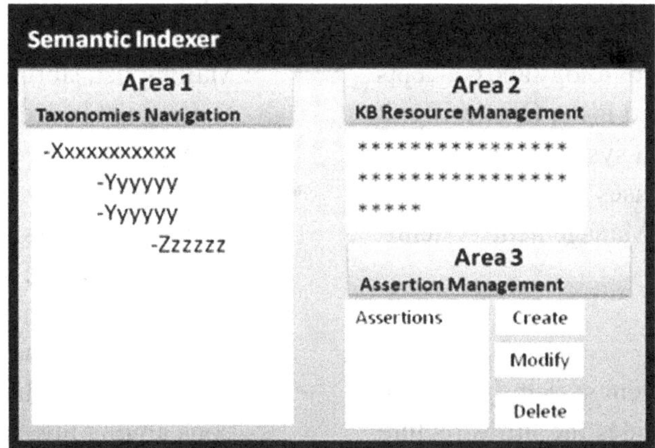

Figure 5. The Semantic Navigator Interface

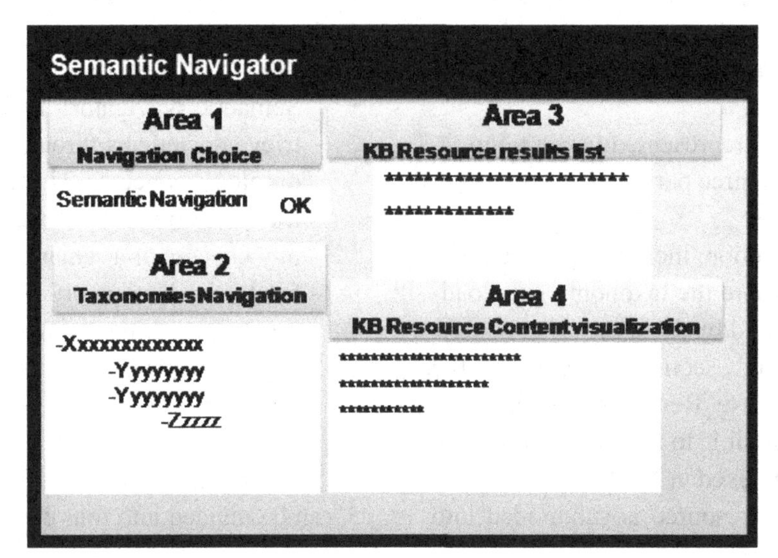

- Navigator choice; in which it is possible to choose the type of navigation; syntactic in this case;
- Syntactic search; in which the user can insert the query string that will activate the research process inside the knowledge base;
- KB resource results list; in which the retrieved resources are listed;
- KB resource content visualization; in which it is possible to see the content of the document selected from the resource list.

Database Management System

The database management system used is a software system designed for the creation and management of users, knowledge, ontology and concepts databases. The DBMS used is the Oracle 10g; it is one of the most efficient systems dedicated to the management of databases and it is a part of the Relational Data Base Management Systems.

Application Server

The knowledge management system is built on an environment that provides the infrastructure for the development of the applications and components servers in a distributed context. The application server used is the OC4J that is based on the J2EE server. The OC4J is easy-to-use and highly productive for developers.

Collaborative Tools

The knowledge management system also includes collaborative tools that are the basis for the front office data flow. Each time a new resource is released, it is converted into XML format in order to index it based on an ontology. The collaborative tools are:

- Email; stores the emails in the knowledge base.
- Mailing list; stores the emails related to the mailing list in order to feed the knowledge base.
- Chat; allows a synchronous interaction among the users. Through the chat, the conversations are saved in order to extract knowledge and make it available in a document format in the knowledge base.
- Forum; allows the asynchronous interactions among the users and stores their par-

Figure 6. The Syntactic Navigator Interface

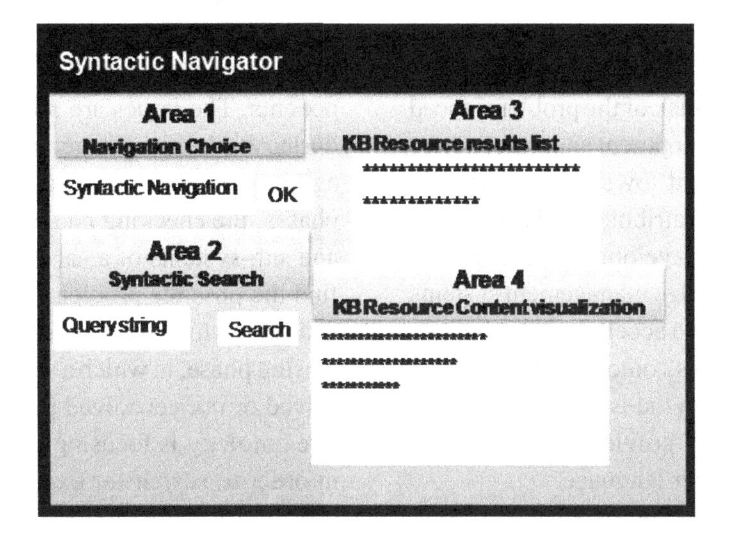

ticipations in order to extract knowledge and make it available in document format in the knowledge base. For that, each participation is saved in XML format, indexed and inserted in the knowledge base.

- News Service; allows the dynamic insertion of news and web pages within the portal. These information will be inserted in a database that manages the pages and the XML representations of the contents in the knowledge base.
- Upload service; allows the users to insert new documents in order to feed the knowledge base.

THE CONTEXT

Ontology-Based KMS for the Issue Management of a COP of Designers in the Automotive Industry

The automotive industry is one of the most important economies in the world. High-performance automotive companies consistently manage all aspects of their business to acquire competitive advantage in the industry. The most efficient plants focused on product development, as new products were emerging at an increasing rate in the market to satisfy customer needs. In order to accelerate the product development, most automakers have adopted processes and technologies that get new products to market faster. To get competitive advantage, automakers apply research and development and customer-centric approaches to every phase of the product development and redesign. Innovation is and will continue to be the necessary element to survive in the dynamic and complex automotive environment.

Moreover, the adoption of technologies is occurring in a fast growth rate in multiple levels in the automotive industry; for example, the IT-based testing of product designs, and the continued automation of the manufacturing facilities. Thus, the integration of technologies in the business is primordial for acquiring competitive advantage in the environment; Technologies' integration in the business provides differentiation to the organization. By leveraging IT-based collaboration, the systems, customers and organizations across the automotive market are connected and they remove the barriers to innovation and competition in the automotive industry.

The issue management is used by firms in the development of new products. It is the process of managing issues through information technologies in order to keep a track of the problems faced during the product development and to document them. Issue management lowers redundancy in problem-solving and contributes to the acceleration of the new product development process. The use of IT-based knowledge management systems is an advanced method to accelerate the management of issues. Besides, ontology-based KMS will be more efficient to the issues management since the ontology will provide the conversion from human to computer language.

In the process of new product development, various multidisciplinary fields are needed, thus many people are involved to release a product providing competitive advantage. In the design phase of a new product, several expertise are playing a major role for the creation of the right designs satisfying the requirements of the conceptualization. The design phase is very important for the new product development effectiveness and is the basis for the rest of the new product development process.

The community of practice, subject of the study, is involved in the design phase of new products within a research center. These latter belongs to a large manufacturing company in the automotive industry. The designers within the research center use a common language, share the same practices and have the same interests which are to achieve optimal products designs. Thus, considered as a community of practice, they communicate and share knowledge concerning the design phase. For the optimization of the CoP activities, the management of the issues faced during the design phase of the complex automotive products is necessary. For that, the community of designers needs a KMS that will achieve this objective. The ontology-based knowledge management system, presented in the previous section, is customized and contextualized to satisfy the needs of the community of designers in the research center.

The developed ontology is related to the automotive context and more precisely to the issues faced during the design of the automotive components. The issues are all the anomalies found during the design of a new product. The issue management of the design phase is divided into three phases; the checking phase, in which the system and sub-systems of a car is analyzed in order to find the possible problems; the solving-problem phase, in which the problem found is solved; the closing phase, in which a statement is made of the solved or not yet solved problems. The automotive ontology is focusing on the checking phase in order to search for existing similar problems and re-use their solutions, if previously solved, otherwise the solutions, if found, are stored on the knowledge base of the KMS. The automotive ontology organizes the automotive knowledge and the users accesses it through the KMS.

The knowledge base consists of documents about the automotive domain. These documents are related to the automotive domain in general and in particular to the issues faced during the new product development activities. The documents include:

- Automotive components documents; including information about the automotive components.
- Issues' cards; including information about the issues faced during the development of a new product.
- Book of edge; an electronic file supporting the issues-solving activities, that register synchronously the issues and their characteristics.
- Issue management documents; released during the various activities related to the issues' management.

The first three types of documents (figure 7) are in excel format and are converted to RDF/RDFS format in order to facilitate the creation of the ontology. The RDF/RDFS files are imported into the Ontology Manager in order to create the automotive ontology; This latter is continuously developed and updated either manually or semi-automatically. The automotive ontology includes concepts and relations that provide information about the product components, the issues and the possible solutions. The issue management documents are released by the members of the community. The documents produced through the collaborative tools are also considered as relevant sources of knowledge. Thus, they are stored in the repository –Knowledge Contents- of the information system, after conversion to XML format. The documents, extracted from the collaborative tools and the issue management documents (figure 7), and the automotive ontology are used by the Indexer in order to create the semantic associations and provide an ontological context to the documents. The semantic associations are then stored in the 'Knowledge Docs' that is the directory. The search engine uses the associations in order to make either the semantic or syntactic navigation.

The technological architecture of the ontology-based KMS, described in the previous section, is adapted to the automotive context. A graphic user interface (GUI) is used by the members of the designers' community, after authentication, to access a customized home page, either for documents insertion or searching. The GUI allows the access to all the functionalities of the knowledge management systems; the knowledge grabber GUI provides the adequate system to create, manage and access the knowledge base that is including the documents related to the issues of the cars; The ontology manager GUI is using the RDFS and OWL language to ensure an efficient development and management of the automotive ontology; The indexer GUI provides an interface to the user in order to insert semantic statements that characterizes the documents related to the automotive issues and that links the statement with the document. And finally, the search engine is used to search and retrieve a document through a semantic or a syntactic research to extract the needed automotive knowledge.

DISCUSSION AND CONCLUSION

The ontology-based KMS contributes to the management of the data related to the automotive context and more precisely to the issue management of the community of designers. The data are stored, organized and reused in an efficient

Figure 7. The Data Flow of the Knowledge Management System

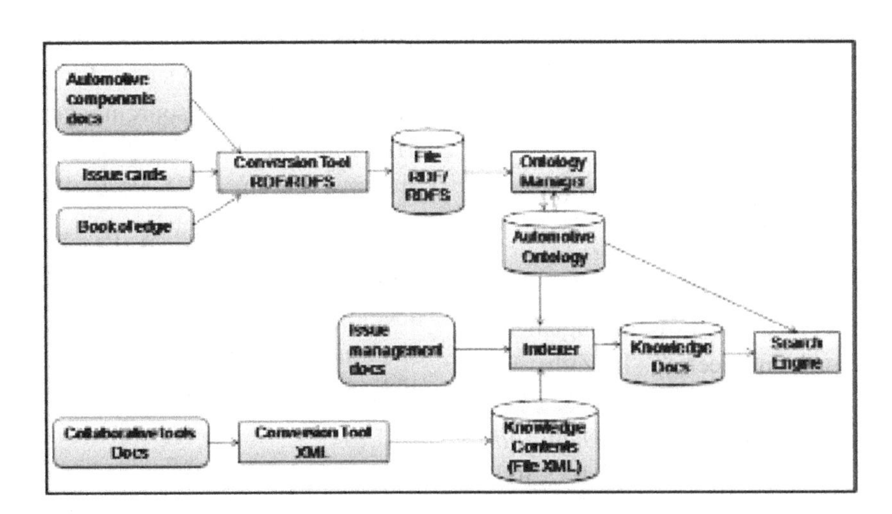

manner to reduce time and redundancy of the design phase of the new product development. The ontology-based KMS provides synchronous and asynchronous channels of communication to support the issue management and the community of designers communicates more easily and develop better the designs of the complex products components.

This new initiative derived from the KIWI project and applied in a complex and dynamic industry as the automotive industry is innovative and allowing the connection among the knowledge management systems, new product development process and communities of practice. The contribution of the combination of technological tools and organizational structure to the new product development process is clear. Through this, the new product development process is optimized and the firms reach high economic performance and competitive advantage in dynamic and complex industries.

REFERENCES

Alavi, M., & Leidner, D. E. (2001). Knowledge management and knowledge management systems: Conceptual foundations and research issues. *Management Information Systems Quarterly*, *25*(1), 107–136. doi:10.2307/3250961

Benbya, H. (2008). *Knowledge management systems implementation: Lessons from the Silicon Valley*. Oxford, UK: Chandos Publishing.

Bernes-Lee, T. (2000). *What the Semantic Web can represent*. Retrieved from http://www.w3.org/DesignIssues/RDFnot.html

Carcagni, A., Corallo, A., Zilli, A., Ingraffia, N., & Sorace, S. (2008). A workflow management system for ontology engineering. In Zilli, A., Damiani, E., Ceravolo, P., Corallo, A., & Elia, G. (Eds.), *Semantic knowledge management: An ontology-based framework*. Hershey, PA: IGI Global.

Cisternino, V., Campi, E., Corallo, A., Taifi, N., & Zilli, A. (2008). Ontology-based knowledge management systems for the new product development acceleration: Case of a community of designers of automotives. *IEEE Proceedings of the 1st KARE Workshop in the 4th SITIS*, Nov. 30- Dec. 3, 2008, Bali, Indonesia.

Colquhoun, G. J., Baines, R. W., & Crosseley, R. (1993). A state of the art review of IDEF0. *International Journal of Computer Integrated Manufacturing*, *6*(4), 252–264. doi:10.1080/09511929308944576

Cooper, R. G. (1994). Perspectives: Third-generation new product processes. *Journal of Product Innovation Management*, *11*, 3–14. doi:10.1016/0737-6782(94)90115-5

Corallo, A., Elia, G., & Zilli, A. (2005). Enhancing communities of practice: An ontological approach. *Proceedings of the 11th International Conference on Industrial Engineering and Engineering Management*, 23-25 April, Northeastern University, Shenyang, China.

Corallo, A., Laubacher, R., Margherita, A., & Turrisi, G. (2009). Enhancing product development through knowledge-based engineering (KBE): A case study in the aerospace industry. *Journal of Manufacturing Technology Management*, *20*(8), 1070–1083. doi:10.1108/17410380910997218

Corallo, A., Lazoi, M., Taifi, N., & Passiante, G. (2010). Integrated systems for product design: The move toward outsourcing. *Proceedings of the IADIS International Conference on Information Systems*, March, 18-21, Porto, Portugal.

Corallo, A., Taifi, N., & Passiante, G. (2008). Strategic and managerial ties of the new product development. *CCIS Proceedings of the First World Summit of the Knowledge Society, 19*, (pp. 398-405). Sept. 24-26, Athens, Greece.

Damiani, E., Ceravolo, P., Corallo, A., Elia, G., & Zilli, A. (2009). KIWI: A framework for enabling semantic knowledge management. In Zilli, A., Damiani, E., Ceravolo, P., Corallo, A., & Elia, G. (Eds.), *Semantic knowledge management: An ontology-based framework*. Hershey, PA: IGI Global.

Davenport, T., & Grover, V. (2001). Knowledge management. *Journal of Management Information Systems*, *18*(1), 3–4.

Füller, J., Bartl, M., Ernst, H., & Mühlbacher, H. (2006). Community-based innovation: How to integrate members of virtual communities into new product development. *Electronic Commerce Research*, *6*, 57–73. doi:10.1007/s10660-006-5988-7

Germain, R. (1996). The role of context and structure in radical and incremental logistics innovation adoption. *Journal of Business Research*, *35*, 117–127. doi:10.1016/0148-2963(95)00053-4

Grant, R. (1996). Towards a knowledge based theory of the firm. *Strategic Management Journal*, *17*(1), 109–122.

Guarino, N. (1997). Understanding building and using ontologies: A commentary to using explicit ontologies in KBS development by van Heijst, Shreiber, & Wielinga. *International Journal of Human-Computer Studies*, *46*, 293–310. doi:10.1006/ijhc.1996.0091

Hopper, M. D. (1990). Rattling sabre-new ways to compete on information. *Harvard Business Review*, *68*(3), 118–125.

Maier, R. (2007). *Knowledge management systems: Information and communication technologies for knowledge management* (3rd ed.). Berlin, Germany: Springer.

March, J. G. (1991). Exploration and exploitation in organizational learning. *Organization Science*, *2*(1), 71–87. doi:10.1287/orsc.2.1.71

Mayer, R. J., Painter, M. K., & deWitte, P. S. (1992). *IDEF family of methods for concurrent engineering and business re-engineering applications*. Knowledge-Based Systems, Inc. Retrieved from http://www.idef.com/pdf/IDEFFAMI.pdf

Nonaka, I. (1991). The knowledge creating company. *Harvard Business Review*, *69*, 96–104.

Nonaka, I., & Takeuchi, H. (1995). *The knowledge creating company: How Japanese companies create the dynamics of innovation*. New York, NY: Oxford University Press.

Sciulli, L. M. (1998). How organizational structure influences success in various types of innovation. *Journal of Retail Banking Services*, *20*(1), 13–18.

Sivadas, E., & Dwyer, F. R. (2000). An examination of organizational factors influencing new product success in internal and alliance based processes. *Journal of Marketing*, *64*(1), 31–50. doi:10.1509/jmkg.64.1.31.17985

Taifi, N., Corallo, A., & Passiante, G. (2008). The strategic orientation of the managerial ties for the new product development. *International Journal of Knowledge and Learning*, *4*(6), 613–624. doi:10.1504/IJKL.2008.022892

Wheelwright, S. C., & Clark, K. B. (1992). *Revolutionizing product development*. New York, NY: Free Press.

Wiig, K. M. (1995). *Knowledge management foundations-thinking about thinking-how people and organizations create, represent, and use knowledge*. Arlington, TX: Schema Press.

Chapter 9
Locating Doctors Using Social and Semantic Web Technologies:
The MedFinder Approach

Alejandro Rodríguez-González
Universidad Carlos III de Madrid, Spain

Ángel García-Crespo
Universidad Carlos III de Madrid, Spain

Ricardo Colomo-Palacios
Universidad Carlos III de Madrid, Spain

José Emilio Labra-Gayo
Universidad de Oviedo, Spain

Juan Miguel Gómez Berbís
Universidad Carlos III de Madrid, Spain

ABSTRACT

The advent of the information age represents both a challenge and an opportunity for medicine. New forms of diagnosis, innovation-oriented supervision and expert location paths are deeply impacting medical sciences as we know it around the word. In this new scenario, semantic technologies can be seen as new and promising tool to support knowledge-based services, and particularly for the health domain, medical diagnosis. This chapter presents MedFinder, a system based on semantic technologies and social Web to improve patient care for medical diagnosis. The main breakthroughs of MedFinder are the follow-up once the diagnosis is performed, by using a medical ontology and formal reasoning together with rules, since it makes possible to locate the most appropriate doctor for a patient using Geographical Information Systems (GIS) and taking into account user preferences given via social Web feedback.

DOI: 10.4018/978-1-61520-921-7.ch009

INTRODUCTION

Internet is a new means for communication, but also a powerful vehicle to match supply and demand of products and services. A shift in the Web content consumer-producer paradigm is making the Web a space for conversation, cooperation and mass empowerment. Emerging killer applications combine sharing information, social dimension, undermining the very principles where content have relied for decades, namely information asymmetry and top-down content delivery. Web 2.0 technologies as outlined in (Laudon & Laudon, 2006) are exemplified by blogs, namely easy to update websites about a particular subject where entries are written in chronological order, picture-sharing environments such as Flickr or Photobucket, social bookmarking sites such as Del.icio.us, video-sharing such as YouTube or music preferences such as Last FM. Web 2.0, social software, social computing, online communities, peer networking etc. In industry, the potential of using Web 2.0 technologies appears to be being gradually realized though adoption (Du & Wagner, 2006) and the medical industry has not been left aside from this process.

According to Giustini (2006), Web 2.0 is changing Medicine as we know it. Several authors have identified tools that highlight the usefulness of the social web as a modifying element for future medical practice and education (McLean et al., 2007). In the education field, many works have been produced that recommend the usage of social web tools for freshmen education (Kamel-Boulos & Wheeler, 2007; Sandars & Schroter, 2007). In the working domain, the utility of Web 2.0 has been demonstrated in professional organization and collaboration (Schleyer et al., 2008; Eysenbach, 2008), volunteer recruitment (Seeman, 2008), patient self care and self-empowerment and research (Seeman, 2008), among other application areas. On the other hand, some researchers have also identified threats regarding the usage of Web 2.0 in the medical environment, such as unwanted behaviors in patients – for example, the avoidance of consultations with physicians (Hughes et al., 2008) or inaccurate online information (Seeman, 2008; Hughes et al., 2008). In relation to the latter case, some initiatives have recently emerged to approve medical web content, using both supervised and automatic methods (Karkaletsis et al., 2008). To sum up, as discussed in Lozerau and Potter (2009), certain technologies such as wikis, blogs or podcasts offer a way to enhance clinical digital learning experiences. Whether these technologies will make medical education and information distribution undergo a revolution will only be known in hindsight (McLean et al., 2007)

Recently, two new terms have come into the arena: Medicine 2.0 and Health 2.0. According to Hughes et al. (2008), Medicine 2.0 is the broader concept and umbrella term which includes consumer-directed "medicine" or Health 2.0. Medicine 2.0 is the use of a specific set of Web tools (blogs, podcasts, tagging, wikis, etc) by entities in health care including doctors, patients, and scientists, using principles of open source and generation of content by users, and the power of networks in order to personalize health care, collaborate, and promote health education.

This scenario has motivated the objective of MedFinder. This tool combines medical diagnosis system which uses on the one hand, semantic technologies, probabilistic techniques and Web 2.0 and on the other hand, geolocation methods, to develop a system to locate the most appropriate doctor for a patient taking into account the context of the patient and the competences of the doctor.

The remainder of the chapter is organized as follows. Section 2 presents the state of the art in similar and related technologies. Section 3 discusses the main features of the approach, the MedFinder conceptual model, together with the architecture. Finally, the paper is completed in Section 4 where conclusions and future work are discussed.

STATE OF THE ART

The aim of this section is to provide a detailed state-of-the-art of the technologies present in this work and a description of the application domain.

LBS, context aware and Geographic Information Systems

Context-awareness is often seen as recently emerged research field within information technology, and in particular within the domain of the Web (Ceri et al., 2007). The term "location-based services" (LBS) is a rather recent concept that integrates geographic location with the general notion of services (Schiller & Voisard, 2004). The origin of the context-aware computing dates back to 1994 in the works of Schilit and Heimer (1994). They suggested the three parameters that determine a context: the location at which the software is used, the collection of nearby people and objects to it and changes of these objects in time. Six years later, Dey and Abowd (2000) proposed a more generic definition: Context is any information that can be used to characterize the situation of an entity. An entity is defined as a person, place or object that is considered relevant for the interaction between a user and an application or between users and applications.

On the other hand, in recent years, medical organizations are devoting increasing attention to Geographical Information Systems (GIS) as a means for sharing and processing geodata of interest (Paolino, Sebillo & Cringoli, 2005). The applications can be both general-purpose, from geodata sharing and dissemination to query formulation, and specific-purpose, such as on-line data processing and location-based services services (LBS) (Peng & Tsou, 2003). Real applications of GIS in medicine are vast (E.g. Avilés et al., 2008; Brewer, 2006; Leslie et al., 2007; Wang & Luo, 2005).

Possibly, in the field of LBS and GIS, the most similar application in relation to MedFinder is the one developed by (Faria et al., 2008), which propose a specific type of monitoring for old people or people with mental illnesses such as Alzheimers, which using a mobile system ensures that the person can be geographically located in any place in the case that he needs help, or physically reached by someone else by means of a button which is designed to deal with such cases.

Medical Diagnosis Decision Support Systems

The diagnosis can be defined as the "description of a health problem in terms of known diseases" and the diagnosis process as a "set of actions needed to obtain the diagnosis" (Van Bemmel & Musen, 2007). On the other hand, a decision support system can provide a set of solutions that best fit the user, depending on different factors concerning the user, the objective or the context where it is applied. Thus, Medical Diagnosis Decision Support Systems (MDSS) (Mangiameli et al., 2004) have become a valuable aid in improving the accuracy of medical diagnosis (e.g. Miller, 1994; Sheng, 2004). The main concept of the MDSS is an inductive engine that learns the decision characteristics of the diseases and can then be used to diagnose future patients with uncertain disease states (Übeyli, 2007).

According to Juarez et al. (2007), the most important problem of these obstacles is, without doubt, the domain complexity, which makes the modeling difficult, and requires advanced techniques to integrate behaviour models and domain descriptions. However, in spite of this known problems, many techniques have been used in medical diagnosis including neuro-fuzzy methods (E.g. Polat & Güneş, 2007; Subasi, 2007), genetic algorithms (E.g. Ahn & Kim, 2009; Yan et al., 2008) or fuzzy inference systems (E.g. Alayón et al., 2007; Tan et al., 2008) to cite the most significant cases. The joint use of semantic web technologies, GIS, LBS and Web 2.0 offers new horizons in MDSS.

Semantic Web

The Semantic Web vision has evolved in recent years as a blueprint for a knowledge-based framework. Thus, the Semantic Web is seen as a promising context for knowledge and data engineering (Vossen, Lytras & Koudas, 2007), and according to Lytras and García (2008), the adaptation of Semantic Web in industrial environments, such as medical, is closer. A proof of this asseveration is the fact that nowadays several initiatives have put together medicine and semantic web.

One way of integration is the use of ontologies in medical domain. Ontologies were developed in the field of Artificial Intelligence to facilitate knowledge sharing and reuse (Fensel et al., 2001). An ontology can be defined as "a formal and explicit specification of a shared conceptualisation" (Studer, Benjamins & Fensel, 1998). According to Sicilia et al. (2009), modern formal ontology eases the creation of knowledge based systems as those required for managing medical information. A recent review on the use of ontologies in Medical domain can be found in Simonet et al. (2008) and Schulz et al. (2009). Real-world examples of the use of Semantic Web in medical diagnosis can be found in many recent and relevant works (E.g. García-Crespo et al, 2010a; García-Crespo et al, 2010b; Hussain et al., 2007; Fuentes-Lorenzo, Morato & Gómez-Berbís, 2009; Podgorelec, Grasic & Pavlic, 2009; García-Sánchez et al., 2008) to cite one of the most relevant and recent cases.

However, Web 2.0 capacities can be extended by the use of Semantic Web technologies. Its usefulness has been identified in medical environments from a generic point of view (Giustini, 2007) as well as in its application (Karkaletsis et al., 2008; Falkman et al., 2008). Not in vain, according to Gruber (2008), the challenge for the next generation of the Social and Semantic Webs is to find the right match between what is put online and methods for doing useful reasoning with the data.

According to the presented state of the art, MedFinder solution gives the opportunity to find in an accurate way to, given a concrete diseases diagnosed to a concrete patient, find more than one physician that will be able to treat the patient. This, is not an easy task, because depending of the diagnosed illness, the patient will need a physician that will be able to manage and to treat, not a concrete disease, a concrete group of diseases or specialty, where the diagnosed diseases will be in.

Once some doctors have been located, the user will be able to choose a concrete physician according to their preferences, which generally will be formed by two parameters:

1. Competence of the doctor (established by a Web 2.0 collaborative ranking)
2. Location of the doctor

When the patient had been treated by the physician, it will be able to score the doctor with some parameters like treatment, consult, accuracy, wait time...

These parameters will modify the score of the physician in order that new users can observe the comments and points given by other users.

Also physicians can score their colleagues taking into account some other parameters like for example research. This will help to other doctors to locate colleagues that work in the same research fields and if they are searching a collaborator find the more suitable according to their preferences.

MEDFINDER

The choice of a medical professional by an expert system to cure or treat a specific disease within a particular field is not a trivial subject. Usually, a person in urgent need of medical treatment arrives at the emergency services of a hospital where a qualified medical professional is assigned to address the problem. However, sometimes the patient is able to select the doctor due to circumstances such as having access to private healthcare. One of the objectives of the current system is to construct a link between the patient and doctor so that the patient can access the professional aptitudes of diverse doctors as a function of his/her necessities, taking into account particular variables, such as the distance which separates both.

An alternative use case scenario for MedFinder is using this tool to hire medical specialists for private or public clinics. Applied to this particular real-world requirements driven scenario, MedFinder needs to be customized. Thus, the component in charge of diagnosis will be no longer needed. In contrast, some new features must be included, mainly derived from Human Resources management issues. In this new environment, the hiring process starts when a concrete need is detected. Once the specialty is entered in MedFinder, the system seeks for available doctors that match needed specialty as well as other criteria, such as user scores or doctor

Description

Given the proposed scenarios and the problems to resolve, the subsequent step is to describe the approximation for the solution of the problems. This solution is based on previous work described and developed in another existing solution, in particular, (García-Crespo, et al, 2010a), which is an existing medical differential diagnosis system based on an ontology, logical inference and probabilistic techniques. Additionally, in what follows, the remainder of the subsystems and parts involved in the elaboration of the framework will be explained.

Architecture

The current structure of MedFinder is based on the construction of various components which interchange certain information among each other. The conjunction of these systems permits that the entire set of components functions correctly, and the necessary data can be obtained to achieve the desired outcome.

A diagram is shown below which displays the general structure of the system and the relationships between its elements. In the subsequent section, the internal functioning of each of the elements will be described, as well as their behavior with regard to the information interchanged with the rest of the components in the system. As additional detail for the understanding of the diagram, consider the following elements:

- *[I]*: Represents the information which one subsystem sends to another.
- *[A]*: Represents the action which one component performs on another component. This may include the transfer of information or not.

ODDX

The first component of the system with which the user interacts is ODDx (García-Crespo et al, 2010a), as can be observed in the diagram. Viewed globally, it is in fact the initial and only part which involves user interaction. It is also this component which is the front end of the expert system that realizes the diagnostic for the user. The fundamental difference of this system compared with the previous system is that the architecture is extended considerably when compared with the existing architecture.

ODDx is an expert system which was capable of inferring medical diagnoses based on a given number of parameters, without any further functions. However, in the new architecture, this component is no more than the beginning of an entire process which not only diagnoses the illness, but also provides the user with much more information related to the diagnosis.

Therefore, the architecture of ODDx has not changed, however, a number of variations have

Figure 1. MedFinder Architecture

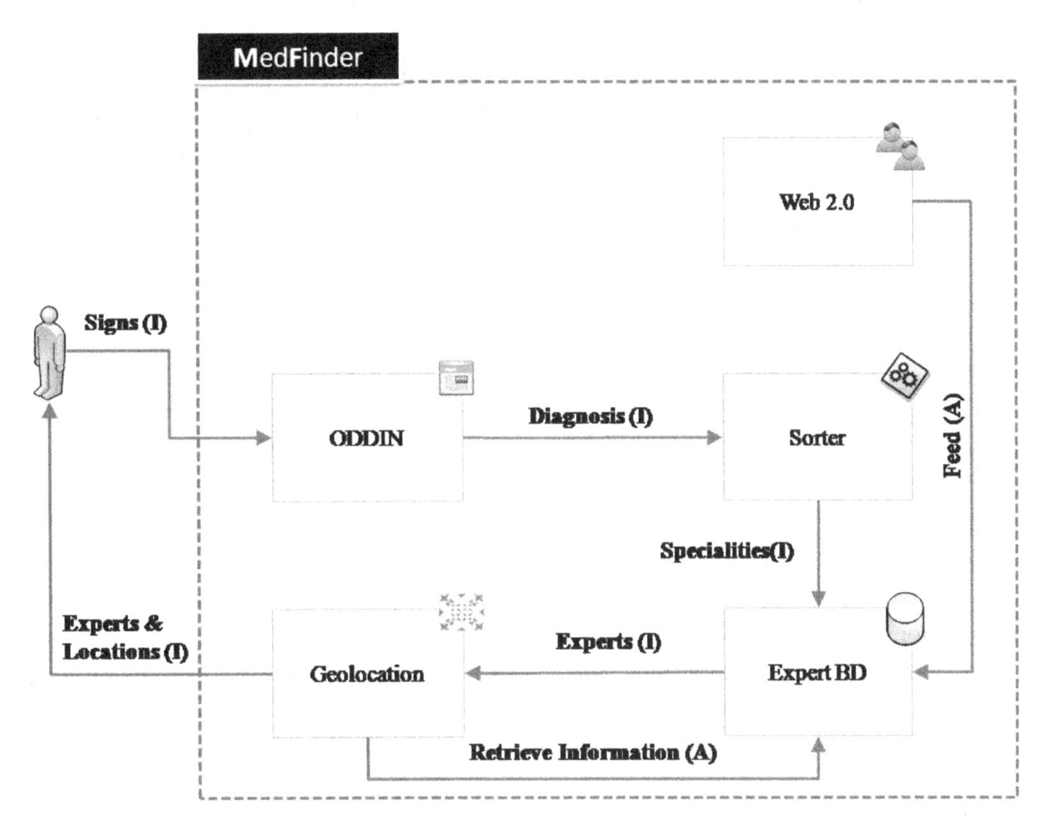

been made. The basic components of ODDx are the following:

- **Probabilistic**: The probabilistic system is the system that is responsible for managing and/or calculating the probabilities of every diagnostic inferred. Every disease that is diagnosed (one or more) from a group of indications has its own probability of happening.
- **Data loading**: This is the engine which performs data loading from the ontology. It employs the Jena API to read the ontology file, which is an open source Java programming environment for Semantic Web applications, and supports the use of various languages, such as RDF, RDFS, OWL and SPARQL.
- **Combination system**: The combination system computes all of the diagnostic combinations possible which may be the result of the interaction of drugs. Basically it is a method which allows, given a patient with a group of indications and associated drugs, the calculation of the possible interactions which may be caused by drugs.
- **Inference**: The inference engine is the main engine of application, because it is a principal constituent of the system which really enables the diagnoses to be made. This engine requires access to the knowledge base with the diseases, symptoms, etc., and at the same time needs access to the knowledge base that contains inference rules.
- **Search**: This component takes the form of a search engine, which realizes SPARQL

queries to the ontology to consult all of the data stored it. This permits fast access to all of the data stored within the ontology.

Sorter

The second component involved in the entire process has been entitled Sorter. This is essentially a system which is capable of generalizing or specifying. It receives an illness as input, which is assigned an ICD code (WHO, 2009), and by means of this code, the system is capable of classifying this illness within one of the multiple specialization of the ICD own classifications.

The implementation of this part of the system may be viewed as both a client and server at the same time. On the one side it acts as a server to receive information (the ICD code of the illness), which is provided by the client and at the same time behaves as a client at the moment it sends its output (the specialization) to the subsequent subsystem.

Whether the use of this subsystem is dependent upon the final implementation is optional. This is because ODDX in its current state does not implement this classification component, which would be easily implementable and in fact a more efficient solution, given that this service could be discarded, thus speeding up the final solution.

Regarding the technologies used to implement this facet, the principal constituent proposed is an ontology, which allows the classification of the illness and its categorization within a specialty. For the present framework, it is proposed that the system uses the same ontology which the ODDX system uses, given that this ontology classifies the illnesses in distinct super classes. In a specific level of these super classes, they represent precisely the specialization. Due to the design of the ontology, the classification of the illness is thus a rather simple process.

Expert Database

The objective of the database of experts is precisely to provide a database which contains the most relevant data relating to an expert. The database contains a series of data, which are listed below:

- **Latitude Coordinate**: Coordinate to locate the expert, principal location.
- **Longitude Coordinate:** Coordinate to locate the expert, principal location
- **Expert Specialization Code:** Code(s) representing the distinct specializations of the expert.
- **Possible Alternative Coordinates:** Reference to possible alternative codes to locate the expert. These coordinates could also be linked to a particular specialization.
- **Patient Scores**: Scores which the patients can assign to the expert once having received the treatment. Further details will be provided below.
- **Expert Scores**: Scores of other experts in the same matter.
- **Other data**: Other series of data such as name, usual address, telephone, email, treatment times, and observations, among others.

An Entity Relationship Diagram is provided in Figure 2, followed by a description of the relations.

Entities:

- **Doctor** – the main entity containing information about medicine doctors, such as name, address, contact and other data.
- **Patient Points** – Patient Point entity represent each mark given by patient to score a doctor.
- **Expert Points** – Export Point entity, similar to Patient Points, is a representation of scores given by doctors/experts to other doctors.

Figure 2. Expert BD E-R Diagram

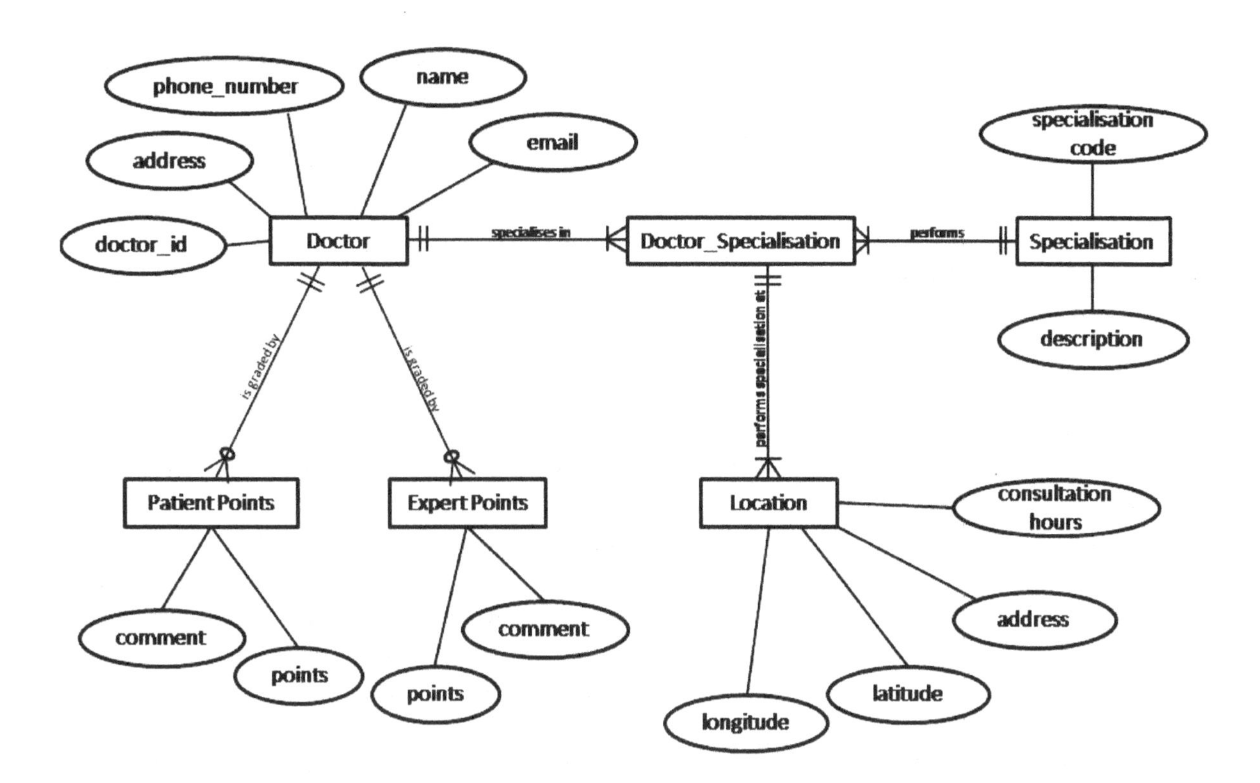

- **Location** – Exact location of the doctor's/expert's office. Contains the exact geolocation data (longitude and latitude), address and hours of consultation.
- **Specialization** – Represents all available experts' specializations, which are described by unified specialization code.
- **Doctor_Specialization** – additional entity which represent each doctor's/expert's specialization. As one expert may perform more than one specialization, each in different location. This entity uses foreign keys from Doctor and Specialization entities.

Relations:

- Specializes_in (related entities: Doctor, Doctor_Specialization) – an one-to-many relation describing doctor's specializations. Since some doctors/experts may perform more than one specialization, each one is represented by a distinct entry in Doctor_Specialization entity.
- Performs_specialization_at (related entities: Location, Doctor_Specialization) – an one-to-many relation defining the locations (office, clinic) where certain doctor's specialization is performed.
- Performs (related entities: Specialization, Doctor_Specialization) – an one-to-many relation describing the specializations that are performed by doctors/experts. It allows to map exact specializations to actual doctors' specializations.
- Is_graded_by (related entities: Patient_Points, Doctor) – an one-to-many relation between doctor's scores and certain doctor. Each doctor is given many scores from pa-

tients. Relation between Expert_Points and Doctor is analogical.

- Additionally, the system described above receives input from another subsystem, namely Web 2.0. Further details of this system will be provided below.

WEB 2.0 FEEDBACK SYSTEM

The aim of the Web 2.0 system is to insert or update data relating to experts within the system. This system also permits the modification of the scores assigned to distinct experts, thereby facilitating the users of the system in the decision to choose one expert or another. This scoring may be divided into two distinct parts:

- **Scores of previous patients**: These scores are based on the scores which patients that have been previously treated by a particular expert may assign to this expert.
- **Scoring of other experts**: This system is based on permitting other experts to assign scores to their colleagues, thereby improving the global vision of these by structuring it such that their knowledge is considered important and other professionals in the sector value their work. Other important thing is that in some cases these points will allow to another physicians to search doctors that are sharing the same or similar research fields in order to find suitable people to share research.

Geolocation System

The geolocation system is a GIS that is the final constituent of the system. It is fundamentally a system which communicates with the database of experts to obtain certain data required to locate the expert. The essential data are the localization, longitude and latitude coordinates, however, other data may be required. The system should calculate a **real** route (realizable by car or on foot, not the distance between two points), which exists between the patient and expert(s).

There are already a number of tools in existence which achieve this aim, however, in the current framework, particular importance was given to the aspects high potency, ease of management and reliability. These features were selected having in mind that the system not only had to establish the physical location of the specialist, but also had to be capable of establishing a route between the patient's current position and the expert.

The API which permits these characteristics is proprietary of Google, *Google Maps*. This framework provides a library which allows the creation of maps of a determined location, establishing distinct tags, controls, and determining routes between two points. Another feature in favor of selecting this platform is that it allows the drawing up of a physical, political and hybrid view between both places (Gibson & Erle, 2006). This implies that the user has more visual references for the place at which he wants to arrive, and additionally, for intermediate points which he must pass to reach his destination (Muller et al., 2004).

However, the key strength of the framework is that it allows the calculation of the shortest route between two points with the simple process of introducing the coordinates of the user and the expert, which in this case can be obtained from ExpertDB. Additionally, a detailed description of the route is provided indicating specifically the route which the user should take, which turns he should take, and the public transport available in each of the streets on the route (Grabler et al., 2008). It also offers further information about the total length of the journey, and the approximate time it takes to complete the entire journey.

Another very interesting aspect is that this API offers the possibility to select how the journey should be realized, on foot or by car, displaying distinct alternatives according to the option chosen. For example, if the user wants to undertake the journey on foot, the system does not process or take into account restrictions or prohibited ways which would be encountered by a car.

CONCLUSION AND FUTURE WORK

Access to healthcare varies across space because of uneven distributions of healthcare providers and consumers (spatial factors), and also varies among population groups because of their different socioeconomic and demographic characteristics (non-spatial factors) (Wang & Luo, 2004). Given this fact, and following the increasing attention of medical organizations in GIS, MedFinder tries to put patients and medical doctors together using semantic technologies and GIS, trying, thus, to better integrate demand and supply shaping spatial factors.

In this chapter, authors outlined MedFinder, a system for recommendation of medical diagnostics which uses an ontology, logical inference, and in particular, introduces the feature of detection of medication that may interact with other medicines and other signs. MedFinder also includes a framework for geolocation of doctors.

A further research step which is envisaged is to also allow users to introduce other complementary parameters into the system to construct a more specific system that allows the user to obtain more precise results. This will be done by expanding the concept of Context, which in MedFinder is a little restricted to geographic coordinates, including aspects like weather, transportation issues and disabilities, to name just a few.

REFERENCES

Ahn, H., & Kim, K. J. (2009). Global optimization of case-based reasoning for breast cytology diagnosis. *Expert Systems with Applications*, *36*(1), 724–734. doi:10.1016/j.eswa.2007.10.023

Alayón, S., Robertson, R., Warfield, S. K., & Ruiz-Alzola, J. (2007). A fuzzy system for helping medical diagnosis of malformations of cortical development. *Journal of Biomedical Informatics*, *40*(3), 221–235. doi:10.1016/j.jbi.2006.11.002

Avilés, W., Ortega, O., Kuan, G., Coloma, J., & Harris, E. (2008). Quantitative assessment of the benefits of specific information technologies applied to clinical studies in developing countries. *The American Journal of Tropical Medicine and Hygiene*, *78*(2), 311–315.

Brewer, C. A. (2006). Basic mapping principles for visualizing cancer data uUsing Geographic Information Systems (GIS). *American Journal of Preventive Medicine*, *30*(2), S25–S36. doi:10.1016/j.amepre.2005.09.007

Ceri, S., Daniel, F., Matera, M., & Facca, F. M. (2007). Model-driven development of context-aware Web applications. *ACM Transactions on Internet Technology*, *7*(1). doi:10.1145/1189740.1189742

Dey, A. K., & Abowd, G. D. (2000). *Towards a better understanding of context and context-awareness*. In CHI 2000 Workshop on the What, Who, Where, When, and How of Context-Awareness.

Du, H. S., & Wagner, C. (2006). Weblog success: Exploring the role of technology. *International Journal of Human-Computer Studies*, *64*(9), 789–798. doi:10.1016/j.ijhcs.2006.04.002

Eysenbach, G. (2008). Medicine 2.0: Social networking, collaboration, participation, apomediation, and openness. *Journal of Medical Internet Research*, *10*(3), e22. doi:10.2196/jmir.1030

Falkman, G., Gustafsson, M., Jontell, M., & Torgersson, O. (2008). SOMWeb: A Semantic Web-based system for supporting collaboration of distributed medical communities of practice. *Journal of Medical Internet Research*, *10*(3), e25. doi:10.2196/jmir.1059

Faria, S., Fernandes, T. R., & Perdigoto, F. S. (2008). *Mobile Web server for elderly people monitoring*. ISCE 2008, IEEE International Symposium on Consumer Electronics.

Fensel, D., van Harmelen, F., Horrocks, I., McGuinness, D. L., & Patel-Schneider, P. F. (2001). OIL: An ontology infrastructure for the Semantic Web. *IEEE Intelligent Systems*, *16*(2), 38–45. doi:10.1109/5254.920598

Fuentes-Lorenzo, D., Morato, J., & Gómez-Berbís, J. M. (2009). Knowledge management in biomedical libraries: A Semantic Web approach. *Information Systems Frontiers*, *11*(4), 471–480. doi:10.1007/s10796-009-9159-y

García-Crespo, A., Chamizo, J., Colomo-Palacios, R., Mendoza-Cembranos, M. D., & Gómez-Berbís, J. M. (2010). (in press). S-SoDiA: A semantic enabled social diagnosis advisor. *International Journal of Society Systems Science*.

García-Crespo, A., Rodriguez, A., Mencke, M., Gómez-Berbís, J. M., & Colomo Palacios, R. (2010). ODDIN: Ontology-driven differential diagnosis based on logical inference and probabilistic refinements. *Expert Systems with Applications*, *37*(3), 2621–2628. doi:10.1016/j.eswa.2009.08.016

García-Sánchez, F., Fernández-Breis, J. T., Valencia-García, R., Gómez, J. M., & Martínez-Béjar, R. (2008). Combining Semantic Web technologies with multi-agent systems for integrated access to biological resources. *Journal of Biomedical Informatics*, *41*(5), 848–859. doi:10.1016/j.jbi.2008.05.007

Gibson, R., & Schuyler, E. (2006). *Google maps hacks: Tips & tools for geographic searching and remixing* (hacks). Sebastopol, CA: O'Reilly Media Inc.

Giustini, D. (2006). How Web 2.0 is changing medicine. *British Medical Journal*, *333*, 1283–1284. doi:10.1136/bmj.39062.555405.80

Giustini, D. (2007). Web 3.0 and medicine. Make way for the Semantic Web. *British Medical Journal*, *335*, 1273–1274. doi:10.1136/bmj.39428.494236.BE

Grabler, F., Agrawala, M., Sumner, R. W., & Pauly, M. (2008). Automatic generation of tourist maps. *ACM Transactions on Graphics*, *27*(3). doi:10.1145/1360612.1360699

Gruber, T. R. (2008). Collective knowledge systems: Where the social Web meets the Semantic Web. *Web Semantics: Science. Services and Agents on the World Wide Web*, *6*(1), 4–13.

Hughes, B., Joshi, I., & Wareham, J. (2008). Health 2.0 and medicine 2.0: Tensions and controversies in the field. *Journal of Medical Internet Research*, *10*(3), e23. doi:10.2196/jmir.1056

Hussain, S., Raza Abidi, S., & Raza Abidi, S. S. (2007). Semantic Web framework for knowledge-centric clinical decision support systems. In *Artificial Intelligence in medicine, (LNCS 4594)* (pp. 451–455). Berlin/Heidelberg, Germany: Springer. doi:10.1007/978-3-540-73599-1_60

Juarez, J. M., Campos, M., Palma, J., & Marin, R. (2007). Computing context-dependent temporal diagnosis in complex domains. *Expert Systems with Applications*, *35*(3), 991–1010. doi:10.1016/j.eswa.2007.08.054

Kamel-Boulos, M. N., & Wheeler, S. (2007). The emerging Web 2.0 social software: An enabling suite of sociable technologies in health and health care education. *Health Information and Libraries Journal*, *24*, 2–23. doi:10.1111/j.1471-1842.2007.00701.x

Karkaletsis, V., Stamatakis, K., Karmapyperis, P., Vojtech, M. A., Leis, A., & Villarroel, D. (2008). Automating accreditation of medical Web content. *Proceedings of the 18th European Conference on Artificial Intelligence* (ECAI 2008), 5th Prestigious Applications of Intelligent Systems (PAIS 2008), (pp. 688-692).

Laudon, K. C., & Laudon, J. P. (2006). *Management Information Systems: Managing the digital firm* (10th ed.). Upper Saddle River, NJ: Prentice Hall.

Leslie, E., Coffee, N., Frank, L., Owen, N., Bauman, A., & Hugo, G. (2007). Walkability of local communities: Using Geographic Information Systems to objectively assess relevant environmental attributes. *Health & Place, 13*(1), 111–122. doi:10.1016/j.healthplace.2005.11.001

Lozeau, A. M., & Potter, B. (2009). Medical information and the use of emerging technologies. *Wisconsin Medical Journal, 108*(1), 30–34.

Lytras, M. D., & García, R. (2008). Semantic Web applications: A framework for industry and business exploitation-what is needed for the adoption of the Semantic Web from the market and industry. *International Journal of Knowledge and Learning, 4*(1), 93–108. doi:10.1504/IJKL.2008.019739

Mangiameli, P., West, D., & Rampal, R. (2004). Model selection for medical diagnosis decision support systems. *Decision Support Systems, 36*(3), 247–259. doi:10.1016/S0167-9236(02)00143-4

McLean, R., Richards, B. H., & Wardman, J. I. (2007). The effect of Web 2.0 on the future of medical practice and education: Darwikinian evolution or folksonomic revolution? *The Medical Journal of Australia, 187*(3), 174–174.

Miller, R. A. (1994). Medical diagnostic decision support systems–past, present, and future: A threaded bibliography and brief commentary. *Journal of the American Medical Informatics Association, 1*(1), 8–27.

Muller Ross, M., Kyusuk, C., Croke, K. G., & Mensah, E. K. (2004). Geographic Information Systems in public health and medicine. *Journal of Medical Systems, 28*(3), 215–221. doi:10.1023/B:JOMS.0000032972.29060.dd

Paolino, L., Sebillo, M., & Cringoli, G. (2005). Geographical Information Systems and online GIServices for health data sharing and management. *Parassitologia, 47*(1), 171–175.

Peng, Z. R., & Tsou, M. H. (2003). *Internet GIS –distributed GIServices for the Internet and wireless networks*. Wiley.

Podgorelec, V., Grasic, B., & Pavlic, L. (2009). Medical diagnostic process optimization through the semantic integration of data resources. *Computer Methods and Programs in Biomedicine, 95*(2), S55–S67. doi:10.1016/j.cmpb.2009.02.015

Polat, K., & Güneş, S. (2007). An expert system approach based on principal component analysis and adaptive neuro-fuzzy inference system to diagnosis of diabetes disease. *Digital Signal Processing, 17*(4), 702–710. doi:10.1016/j.dsp.2006.09.005

Sandars, J., & Schroter, S. (2007). Web 2.0 technologies for undergraduate and postgraduate medical education: An online survey. *Postgraduate Medical Journal, 83*, 759–762. doi:10.1136/pgmj.2007.063123

Schilit, B. N., & Heimer, M. M. (1994). Disseminating active map information to mobile hosts. *IEEE Network, 8*(5), 22–32. doi:10.1109/65.313011

Schiller, J., & Voisard, A. (2004). *Location-based services*. Amsterdam, The Netherlands: Elsevier.

Schleyer, T., Spallek, H., Butler, B. S., Subramanian, S., Weiss, D., & Poythress, M. S. (2008). Facebook for scientists: Requirements and services for optimizing how scientific collaborations are established. *Journal of Medical Internet Research, 10*(3), e24. doi:10.2196/jmir.1047

Schulz, S., Stenzhorn, H., Boeker, M., & Smith, B. (2009). Strengths and limitations of formal ontologies in the biomedical domain. *Electronic Journal of Communication Information & Innovation in Heath, 3*(1), 31–45.

Seeman, N. (2008). Web 2.0 and chronic illness: New horizons, new opportunities. *Healthcare Quarterly (Toronto, Ont.)*, *11*(1), 104–110.

Sheng, O. R. L. (2004). Editorial: Decision support for healthcare in a new information age. *Decision Support Systems*, *30*(2), 101–103. doi:10.1016/S0167-9236(00)00091-9

Sicilia, J. J., Sicilia, M. A., Sánchez-Alonso, S., García-Barriocanal, E., & Pontikaki, M. (2009). Knowledge representation issues in ontology-based clinical knowledge management systems. *International Journal of Technology Management*, *47*(1/2/3), 191-206.

Simonet, M., Messai, R., Diallo, G., & Simonet, A. (2008). Ontologies in the health field. In Berka, P., Rauch, J., & Abdelkader Zighed, D. (Eds.), *Data mining and medical knowledge management: Cases and applications*. Hershey, PA: IGI Global.

Studer, R., Benjamins, V. R., & Fensel, D. (1998). Knowledge engineering: Principles and methods. *Data & Knowledge Engineering*, *25*(1-2), 161–197. doi:10.1016/S0169-023X(97)00056-6

Subasi, A. (2007). Application of adaptive neuro-fuzzy inference system for epileptic seizure detection using wavelet feature extraction. *Computers in Biology and Medicine*, *37*(2), 227–244. doi:10.1016/j.compbiomed.2005.12.003

Tan, T. Z., Quek, C., Ng, G. S., & Razvi, K. (2008). Ovarian cancer diagnosis with complementary learning fuzzy neural network. *Artificial Intelligence in Medicine*, *43*(3), 207–222. doi:10.1016/j.artmed.2008.04.003

Übeyli, E. D. (2007). Implementing automated diagnostic systems for breast cancer detection. *Expert Systems with Applications*, *33*(4), 1054–1062. doi:10.1016/j.eswa.2006.08.005

Van Bemmel, J. H., & Musen, M. A. (1997). *Handbook of medical informatics*. Springer-Verlag.

Vossen, G., Lytras, M., & Koudas, N. (2007). Editorial: Revisiting the (machine) Semantic Web: The missing layers for the human Semantic Web. *IEEE Transactions on Knowledge and Data Engineering*, *19*(2), 145–148. doi:10.1109/TKDE.2007.30

Wang, F., & Luo, W. (2005). Assessing spatial and nonspatial factors for healthcare access: Towards an integrated approach to defining health professional shortage areas. *Health & Place*, *11*(2), 131–146. doi:10.1016/j.healthplace.2004.02.003

WHO. (2009). *ICD: International statistical classification of diseases and related health problems*. World Health Organization. Retrieved from http://www.who.int/classifications/icd/en/

Yan, H., Zheng, J., Jiang, Y., Peng, C., & Xiao, S. (2008). Selecting critical clinical features for heart diseases diagnosis with a real-coded genetic algorithm. *Applied Soft Computing*, *8*(2), 1105–1111. doi:10.1016/j.asoc.2007.05.017

Chapter 10
Refactoring and its Application to Ontologies

Jordi Conesa
Universitat Oberta de Catalunya, Spain

Antoni Olive
Universitat Politècnica de Catalunya, Spain

Santi Caballé
Universitat Oberta de Catalunya, Spain

ABSTRACT

Over the last years, a great deal of ontologies of many different kinds and describing different domains has been created, and new methods and prototypes have been developed to search them easily. These ontologies may be reused for several tasks, such as for increasing semantic interoperability, improving searching, supporting Information Systems or the creation of their conceptual schemas. Searching an ontology that is relevant to the users' purpose is a big challenge, and when the user is able to find it a new challenge arises: how to adapt the ontology in order to be applied effectively into the problem domain. It is nearly impossible to find an ontology that can be applied as is to a particular problem. Is in that context where ontology refactoring takes special interest. This chapter tries to clarify what ontology refactoring is and presents a possible catalog of ontology refactoring operations.

INTRODUCTION

Lately, lots of ontologies of many different kinds and describing different domains have been created and new methods and tools have been developed to search them easily. Maybe the most prominent of these tools is Swoogle[1], a web-based ontology search engine that searches over ten thousand available ontologies. Even though the facilities these new search technologies provide, finding an ontology that is relevant to a given purpose is a big challenge. And even, when we are able to find an ontology that is good enough to be reused, a new challenge arises: how to adapt the ontology in order to be applied effectively into the problem domain. It is nearly impossible to find an ontology that can be applied as is to a particular problem. In fact, once found an ontology

DOI: 10.4018/978-1-61520-921-7.ch010

it needs to be restructured to make it more usable (by deleting irrelevant terms, adding additional constraints…) and to adapt it to the conceptualization the user has in mind (by restructuring it to represent the same semantics but with a better structure) (Conesa J, 2008).

Evolution operations have played an important role in the software engineering field. Some of these operations aim to modify a program or a conceptual schema and can be used to add, modify or delete the functionalities of a system or to improve the quality of the system without modifying its semantics. These operations are called refactoring operations when they are applied to object-oriented artifacts or restructuring operations when applied to other kind of artifacts. Over the last decade, several researchers have attempted to use refactoring operations not only in software but at higher degrees of abstraction (Sunye, Pollet et al., 2001): databases, Unified Modeling Language (UML) models (OMG, 2003), Object Constraint Language (OCL) rules (OMG, 2003), and more shyly in ontologies. But not only refactoring operations allow for modifying the structure of something while keeping its semantics, other technologies propose similar evolution operations that can be considered as refactoring operations, such as some schema evolution operations of, program slicing operations…

Ontology refactoring is the process of modifying the structure of ontologies but preserving their semantics (Conesa J, 2008). Refactoring has successfully applied to several domains (Mens and Tourwe, 2004), but has not applied effectively to ontologies yet. The two main lacks of the application of refactoring to ontologies is the lack of an unambiguous and agreed definition of ontology refactoring and a catalog of the refactoring operations that makes sense for refactoring ontologies. Such agreed catalogs are present within other fields such as software refactoring[2] and database refactoring[3].

We believe that the refactoring work of other fields, in combination with other techniques such as program restructuring (Griswold and Notkin, 1993) or schema transformation (Batini, Ceri et al., 1992), may be used to successfully define ontology refactoring and identify a catalog of its potential operations within the context of ontologies. Therefore, the goals of this chapter are: 1) to define clearly what a refactoring is, 2) to analyze various techniques for restructuring programs and schemas in order to identify in what cases these techniques or their operations can be useful for ontology refactoring, and 3) to present an exhaustive catalog of ontology refactoring operations that includes, after being adapted, operations from refactoring in other fields (software, databases and UML models mostly) and from other schema transformation techniques.

In the next section, the reader will find a brief history of existing restructuring techniques and how they have evolved to become the refactoring techniques we have today. Section 3 defines ontology refactoring taking into account the definitions of refactoring in other fields. Later, a catalog of ontology refactoring operations is presented. Such a catalog has been created by adapting refactoring operations from other techniques or from the refactoring of other fields. Finally, section 5 concludes the chapter presenting our conclusions and lessons learnt.

FROM PROGRAM REESTRUCTURING TO REFACTORING PASSING THROUGH SCHEMA TRANSFORMATION

In the last decades, several operations with common purposes but different uses have been defined using different names: refactoring operations (Opdyke, 1992; Moore, 1996; Roberts, Brant et al., 1997; Fowler, 1999; Ambler, 2003), restructuring operations (Eick, 1991; Batini, Ceri et al, 1992; Assenova and Johannesson, 1996; Halpin, 2001), program transformation operations (Griswold and Notkin, 1993; Griswold, Chen et al., 1998), model transformation operations (Sunye, Pollet et al., 2001; Roberts, 1999; Correa and Werner, 2004), etc. This section examines chronologically the techniques behind these names in order to show that some of they pursue the same goal: modifying the structure of a given artifact without changing its semantics, while improving some of its quality factors. We will also see that most of these techniques were conceived to restructure programs and therefore they should be adapted if we want to use them to restructure ontologies.

Program Restructuring

Program restructuring (Griswold and Notkin, 1993) is a technique developed in the, 1980s. Its purpose is to restructure programs without modifying their behavior. This technique differs from refactoring in that it only deals with non-object-oriented programs. The following paragraphs discuss the main restructuring techniques that appeared in the 1990s.

Griswold and Notkin (1993) presented a set of restructuring operations that improve the quality of a program while maintaining its behavior. These operations make it possible to change the name of variables, replace expressions with variables or variables with expressions, replace a sequence of commands with a function that executes them and a call to such a function, etc. The same paper showed how to automate the execution of these operations. The same authors also defined a framework that supports the user in the detection and execution of restructuring operations for large programs (Griswold, Chen et al., 1998).

There are other program restructuring approaches, but they are not applicable to ontologies due to the limitations they impose such as in (Bergstein, 1991), where a set of five operations for restructuring an object-oriented schema (or program) while maintaining its possible instances. This kind of preservation is called object-preserving, and means that all non-abstract classes of the two schemas have the same name and attributes, including the inherited ones.

Schema Transformation

Schema transformation (Batini, Ceri et al., 1992) has been widely studied in the last two decades. The following lines describe schema transformation and summarize the most relevant work in this area that should be taken into account in defining ontology refactoring.

Schema transformations are applied to an initial schema and produce a resulting schema. These transformations can be classified based on how they change the information contained in the transformed schema (Batini, Ceri et al., 1992):

- *Information-preserving transformations*: The execution of the transformation preserves the information contained in the schema.
- *Information-changing transformations*: These may be further classified as 1) *augmenting transformations*: The resulting schema contains more information than the original one; 2) *Reducing transformations*: The resulting schema contains less information than the original one; and 3) *Non-comparable transformations*: Neither of the previous categories is true.

Schema transformation operations that belong to the first group may be regarded as refactoring operations because they maintain the semantics of the modified schema, i.e. the information it contains. We are interested in this kind of operation. Hereinafter, when we use the term schema transformation operations we are referring to information-preserving transformation operations.

Identifying whether two conceptual schemas define the same information (i.e. they are equivalent) is very difficult. Several authors have studied this topic and presented different equivalence proposals (Hofstede, Proper et al.; Batini, Ceri et al., 1992; Proper and Halpin, 2004) based on: sets, logics, contextual equivalence, substitution equivalence, etc. We believe that the most semantically coherent definition is:

Two conceptual schemas are equivalent if and only if whatever universe of discourse, state or transition can be modeled in one can also be modeled in the other. (Halpin, 2001)

Unfortunately, it is very difficult, or even impossible, to demonstrate that two schemas are equivalent using this definition. Therefore, most schema transformation approaches use the following definition of equivalence for substitution:

two schemas are equivalent if and only if any of them may be derived from the other (Halpin and Proper, 1995)

From the last definition we can infer that schema transformation operations are bidirectional. Therefore, a schema transformation operation can transform schema S1 to another schema S2 and undo the change by transforming schema S2 to schema S1.

Batini et al. (1992) defined a set of information-preserving schema transformation operations that improve the readability and conciseness of conceptual schemas (Batini, Ceri et al., 1992). These operations are:

1. *Deletion of redundant cycles of relationships*: This operation deletes paths made up of relationship types already defined by other relationship types.
2. *Addition/deletion of derived attributes/subsets*: This operation adds or deletes derived attributes or classes.
3. *Elimination of dangling subentities in generalization hierarchies*: This operation merges an entity with its subtypes. The application is only valid when the subtypes are empty and therefore they do not contain any field.
4. *Elimination of dangling entities*: An entity absorbs another entity related to it by relationship type.
5. *Creation of a generalization*: This operation creates generalization and specialization relationships between entities with shared attributes.
6. *Creation of a new subset*: This operation is executed when an entity E_1 participates in a relationship type with a minimum cardinality of zero. In such a case, a new subtype of E_1 is created that only contains the registers of E_1 that participate in the relationship type. The relationship type is also moved to the new subtype, and its minimum cardinality is changed to one.

Batini et al. (1992) also defined schema transformation operations that preserve the semantics of a transformed schema but which convert it to its third normal form. We consider these operations technology-dependent and therefore not relevant for this chapter.

(Eick, 1991) presented six schema transformation operations that make it possible to:

- Move attributes between the entity types of the conceptual schema related by relationship types and generalization/specialization relationships.
- Split a class in two and relate the two resulting classes using a relationship type.
- Create a new subtype or supertype of an existing entity type.
- Move the ends of the relationship types through the taxonomy of entity types.

(Assenova and Johannesson, 1996) compiled the most relevant information-preserving schema transformation operations and added a new operation to the list. They also analyzed how to apply these operations to improve the quality of a conceptual schema by studying the quality factors improved by the execution of each operation. Eight of the nine operations presented focus on restructuring relationship types, and only one focuses on restructuring entity types. As the authors mention in the paper, the presented list is incomplete and it would be useful to extend it with new operations.

Halpin (2001) defined a set of information-preserving schema transformation operations for ORM (Halpin, 1996) conceptual schemas. We believe that two of his operations are especially relevant to our work:

- Specialize or generalize a predicate: This operation replaces an n-ary relationship type with m relationship types with an arity of n-1. As mentioned above, this operation is bidirectional. Therefore, it also makes it possible to merge m n-ary relationship types into one n+1-ary relationship type.
- Absorb a relationship type: By applying this operation, a relationship type that restricts the possible values of another relationship type is absorbed by the latter.

All relevant operations presented in this section are included in the catalog of ontology refactoring operations of Section 3.

Software Refactoring and its Evolution

The concept of refactoring was born in the early, 1980s when Opdyke (1992) defined refactoring as "*a program transformation that reorganizes a program without changing its behavior*".

Before Opdyke (1992), several authors presented the idea of creating a set of modification operations for object-oriented programs that preserve their behavior. There are some algorithms and heuristics that produce good[4] class organizations. Examples include (Bergstein, 1991), which proposed a set of class-transformation operations that preserve the behavior of an object-oriented schema with certain structural limitations (discussed in the previous subsection); and Casais (1989), which presented a set of techniques for restructuring the inheritance of a class diagram without losing functionality.

In his thesis, Opdyke (1992) defined 23 primitive refactorings and showed how to compose them in order to create more complex refactoring operations. For each refactoring operation, this author defined a precondition that guarantees that the behavior of the program will not change after the operation is executed. Opdyke defined the software refactoring operations in terms of C++.

Later, Roberts (1994) weakened Opdyke's refactoring definition in order to be able to define new refactorings that do not preserve program behavior but may be useful for designers. Roberts also extended the refactoring definition with postconditions in order to simplify the creation of complex refactoring operations by sequencing simpler ones and decrease the amount of analysis needed to identify what complex refactoring chains can be applied.

However, we believe that the best software refactoring definition is the one presented by Fowler (1999):

- **Refactoring (noun):** A change made to the internal structure of software to make it easier to understand and cheaper to modify without changing its observable behavior.
- **To refactor (verb):** To restructure software by applying a series of refactorings without changing its observable behavior.

According to Philipps and Rumpe (2001), the definition of external or observable behavior of a given program depends of its functional requirements. For example, in real-time information systems, execution time is considered a functional requirement. Therefore, in that case many typical refactoring operations may be inapplicable because their changes may negatively affect the execution time and therefore violate the functional requirements and therefore change its observable behaviour.

Fowler (1999), in his book, defines 72 refactorings and briefly summarizes each of them, explaining, for each operation, under what conditions it is useful, its execution steps, and the validations that should be carried out after its execution in order to guarantee that the program's behavior is preserved. Fowler also defined the code smells, which are conditions that denote certain structures in the code that suggest (and sometimes cry out for) the possibility of refactoring. With each code smell, Fowler associates the set of refactoring operations that can be applied to improve the program.

Software refactoring became very important with the release of Extreme Programming (Beck, 1999), a programming technique characterized by repetitive and aggressive application of refactoring operations to a program.

Although refactoring was designed to be applied manually, some tools have been developed to perform it automatically, such as Refactoring Browser (Roberts, Brant et al., 1997), RefactorIT[5], Together ControlCenter[6], etc.

Despite the contribution of the code smells, knowing when and where to apply refactoring operations is still a big challenge. Finding what refactoring operations can be applied effectively and where is what is known as detecting refactoring opportunities. When detecting refactoring opportunities the following questions should be answered:

- What refactoring operations can be applied effectively: the refactoring operations whose execution will improve the program should be obtained.
- Where they can be applied: in what parts of the program the previous refactoring operations can be applied effectively.
- In which order they should be applied: refactoring operations should be sequenced when several refactoring operations can be applied effectively at the same moment. Inferring in what order they should be applied is not easy since it will depend on the quality factor we want to improve and the application of one operation may imply that another operation cannot be applied anymore and therefore should be discarded.
- What refactoring operations should not be applied: since refactoring operations are unidirectional there are operations that undo the modifications done by other operations. This should be taken into account in order to avoid cycles in the execution of refactoring operations.

In order to completely automate the refactoring process, we must:

1. Automatically determine what operations can be applied and to what part of the schema, and
2. Execute them automatically.

As aforesaid, the first step is known as identifying the refactoring opportunities. The second step has been studied intensely in recent years (Roberts, Brant et al., 1997; Roberts, 1999), and is now close to being solved.

The detection of refactoring opportunities is non-deterministic and a lot of semantic information is required to identify the best refactoring operation for each case. Hence, this detection is only possible in some cases, and even in such cases its automation is not trivial.

For all these reasons, the refactoring process cannot be totally automated and require designer/programmer intervention in order to be applied successfully. The most sophisticated tools detect some refactoring opportunities and use them to propose refactorings, but it is usually the programmer who in the last term has to hand-pick what refactoring operations to apply and the order in which they are applied.

Several approaches to automatically selecting and applying certain refactoring operations in certain contexts and cases have appeared (Ducasse, Rieger et al., 1999; Kataoka, Ernst et al., 2001; Simon, Steinbrukner et al., 2001; Gorp, Stenten et al., 2003; Koru, Ma et al., 2003; Tourwe and Mens, 2003; Xing and Stroulia, 2003; Mens and Tourwe, 2004; Rysselberghe and Demeyer, 2004; Grant and Cordy, 2003; Simon, Steinbrukner et al., 2001; Kataoka, Ernst et al., 2001; Tourwe and Mens, 2003; Koru, Ma et al., 2003).

Conceptual Schema Refactoring

Most of current integrated development environments support the refactoring of programs. These tools work well for small granularity refactorings such as the extraction of a fragment of source code to a new method (*extract method*) or the one that renames the member of a class (*rename member*), but they are neither intuitive nor easy to work with for big refactorings such as when a class is split into several classes and its original attributes and methods distributed between them (*extract hierarchy*).

On the other hand, inconsistencies can appear between the source code and its specification. To solve these inconsistencies, refactoring operations can be applied on a higher abstraction level (the level of specification, analysis or design) and such changes automatically propagated to the lower abstraction levels. Several authors have tried to adapt software refactoring to conceptual schemas, such as Sunye et al. (2001), the people of the Refactoring Project group (Gorp, Stenten et al., 2003; Tourwe and Mens, 2003; Mens, 2004 and Porres, 2003).

Some authors use a tool to perform refactoring operations on UML models (Astels, 2002; Marko Boger and Fragemann, 2002). This tool maintains consistency between the UML diagrams and the associated source code. Other authors, such as Bottoni, Parisi-Presicce et al. (2002), have proposed an inverse technique – that is, applying refactoring operations directly to the source code and defining techniques to propagate the changes to the associated conceptual schemas to maintain consistency.

Even though these proposals take conceptual schemas into account in the refactoring process, their vision of a conceptual schema is too simple. They see it as a very low-abstraction-level schema, usually represented by a set of UML models that directly represent a program. Using the MDA approach (Kleppe, Warmer et al., 2003) we can define these models as platform-specific models (PSM).

Table 1 summarizes the main techniques of conceptual schema refactoring. This table describes the purpose of such approaches, the artifacts they refactor, the abstraction level in which they work, the refactoring operations they define and their drawbacks in our context: the need for a refactoring catalog for ontologies.

Roberts (1999) offered a definition for model refactoring:

A model refactoring is a pair R = (pre;T) where pre is the precondition that the model must satisfy, and T is the model transformation.

This definition takes the operational perspective on refactoring. The definition allows certain operations that do not preserve the semantics of a refactored model to be considered refactoring operations.

Sunye et al. (2001) were the first to define the refactoring of UML models by defining a set of refactoring operations applicable to class diagrams and statecharts. However, they only explicitly formalized and defined the refactoring operations that deal with state machines. Their paper was the first to show refactoring as an analysis and design technique rather than a technique that was only applicable to source code. However, the authors did not study how the execution of refactoring operations affects the integrity constraints of the refactored schema, which in the case of the UML

Table 1. Summary of the main approaches to conceptual schema refactoring

Approach	Purpose	Artifact / Abstraction level	Operations defined	Drawbacks
(Sunye, Pollet et al., 2001)	To define the refactoring of UML models	UML class diagrams and state machine diagrams / Design level	*Insert generalization element (GE), Delete GE, Move method, Generalize element* and *Specialize element*	Preconditions and postconditions not fully specified Too simple to be applicable to analysis levels
(Marko Boger and Fragemann, 2002)	To define a tool to refactor UML models	UML state machines, activity and class diagrams / Design level	*Rename method, Move up method, Rename class* and *Merge states*	Not much formalization The available document has little and informal information, and the supposed full document is written in German
(Zhang, Lin et al., 2005)	To define transformation operations at the metamodel level	Any model that can be expressed as an instance of a metamodel: Petri nets, quality models, UML models, etc. / Design level	Subset of Fowler's catalog: *Add class, Extract superclass, Extract class, Remove class, Move class, Rename class, Collapse hierarchy, Add attribute, Remove attribute, Rename attribute, Pull up attribute* and *Push down attribute*	Not focused on defining refactoring operations The defined operations are too simple
(Judson, Carver et al., 2003)	To use patterns to define transformation operations at the metamodel level	Any UML diagram / Any abstraction level	None	Not focused on defining refactoring operations
(Porres, 2003)	To define refactoring operations as transformation rules and show how to automate and define them	Any UML model / Any abstraction level	*Encapsulate attributes, AddSetter and AddGetter.*	It defines how to formalize refactoring operations, but not their catalog or opportunities The user needs to be fully familiar with the UML metamodel
(Gorp, Stenten et al., 2003)	To extend the UML metamodel to link the source code with UML models. This link is used to define and automate two refactoring operations	UML models / Design level	*Extract method* and *Pull up method*	The metamodel extension became useless after the addition of Action Semantics in UML 1.5, which subsumed the proposed metamodel
(Correa and Werner, 2004)	To adapt the concept of refactoring to OCL integrity constraints	UML integrity constraints / Any abstraction level	*Add variable from expression, Split AND expression, Add operation definition from expression* and *Replace expression by operation call*	It does not study how the modification of UML models affects OCL integrity constraints, define an exhaustive list of OCL refactoring operations, or create any new operations (they are all adapted from Fowler's operations)

are expressed using OCL (OMG, 2003). Finally, the authors classified refactoring operations in five groups, based on the kind of action they perform: insertion, deletion, movement, generalization and specialization. For each group, they discuss the following:

- The refactoring operation *Insert generalizable element* adds a new element between two adjacent elements of a taxonomy (parent and child). Their operations apply the *generalizable element* concept to both classes and associations. It is not clear how this can be done and under what conditions it is allowed for associations.
- The operation *Remove generalizable element* is the inverse of the previous operation. It deletes an element without changing the behavior of the taxonomy. The operation also links the subtypes of the deleted element to its supertypes. This operation cannot be executed when the element is directly or indirectly referred to in other diagrams.
- The operation *Move* is used to transfer a method from one class C_1 to another class C_2. After moving the method to C_2, a new method that calls the moved method should be created for C_1. The following conditions must be guaranteed before this operation is applied:
    - A navigable binary association with a multiplicity of one must exist between C_1 and C_2, and
    - The code of the operation to be moved must not refer to the attributes of C_1.
- *Generalization* refactoring may be applied to elements contained in classes, such as attributes, methods, operations and association ends. Since private characteristics are not accessible to the subclasses, they cannot be moved. Before moving an element, all subtypes of the destination class must have defined the characteristic to be moved.
- *Specialization* refactoring is the inverse of the previous operation. It consists in sending an element to the subtypes of the class to which it belongs.

In presenting the above operations, the authors do not formally define their preconditions and the textual conditions they present are incomplete. Furthermore, their paper does not take into account how to propagate refactoring changes to other UML models or to integrity constraints (even graphical ones such as cardinalities, disjointness and completeness), the refactoring opportunities that allow us to determine what refactoring operations may be applied, or the bad smells that indicate when a model should be refactored.

Hence, obtaining a catalog of refactoring operations from the aforementioned paper is difficult. Furthermore, it is unclear whether all the presented operations are refactoring operations, because we believe that some of them are closer to evolution schema operations: *Add attribute, Add method, Delete attribute, Delete class*, and so forth. Other operations can be seen as adaptations of Fowler's refactoring operations, such as the operation *move* which can be seen as an adaptation of the *Move method* operation or the operation *Generalization* which can be seen an adaptation of Fowler's *Pull up method/attribute* operations.

Refactoring presented by Sunye et al. (2001) works at design level, which means that it assumes the UML diagram is a direct representation of a program and therefore avoid using high level abstraction constructs such as ternary associations, multiple inheritance, multiple classification, association classes, etc. Applying refactoring operations on design level has its advantages, but applying them to a higher abstraction level provides even more benefits because an improvement on a higher abstraction level tends to generate *n* improvements at design level, with *n* greater than 2.

Ever though the validity of conceptual schema refactoring was demonstrated, efforts have focused on refactoring automation and in particular on how to represent refactoring operations to support a CASE tool to automate refactoring opportunities and execution. For example, Marko Boger and Fragemann (2002) defined various refactoring operations for static and dynamic UML models and presented a browser that supports the execution of the presented refactoring operations.

In the same direction, Zhang, Lin et al. (2005) presented a framework for automatically executing refactoring operations on models. This framework works with metamodels and can therefore refactor other kinds of models, such as Petri nets, service quality models and so forth. With UML, this framework can execute a subset of the operations defined by Fowler (1999): *Add class, Extract superclass, Extract class, Remove class, Move class, Rename class, Collapse hierarchy, Add attribute, Remove attribute, Rename attribute, Pull up attribute* and *Push down attribute*. Moreover, the tool allows to define new refactoring operations using a language called *Embedded Constraint Language* (ECL) (Gray, 2002), which is an extension of OCL that allows actions on conceptual schemas to be defined. This tool directly adapts software refactoring operations without taking certain model characteristics into account. It therefore has some drawbacks, such as the lack of refactoring operations to deal with association ends (moving up and down associations in a taxonomy for example). In consequence, basic operations such as *Collapse hierarchy* (which merge a class with its subtypes) cannot be applied to a class that participates in associations.

Judson, Carver et al. (2003) defined model transformation operations using metamodels. They created patterns at the metamodel level that graphically define various model transformation families that can be systematized. Although this approach allows us to write refactoring operations for all kinds of UML models, it focuses on representing transformation operations through patterns rather than defining a catalog of refactoring operations.

Porres (2003) proposed implementing conceptual schema refactorings as a collection of transformation rules. Each transformation rule is defined by five elements: its name, a comment, a sequence of parameters, a guard condition that defines when the rule can be applied, and the body of the rule, which implements its effect. Porres' paper also defined a simple algorithm that simulates the intuition of the designer in the application of some refactorings. The algorithm detects the refactoring opportunities and sequences them automatically using some heuristics. The SNW tool (Porres, 2002) implements Porres' approach and allows editing UML models and creating, deleting and modifying refactoring operations.

There is a very well-known research group at the University of Antwerp called *Refactoring Project*[7] very prominent in the area of model refactoring. Its main goal is to design tools that make refactoring easier, more usable and more automatic. The group uses graph-rewriting algorithms to automate refactorings. Specifically, models are rewritten as graphs and then tools of graph management are used to detect refactoring operations that can be applied. One problem of such an approach is that the graphs of models are too large and therefore it is quite difficult for designers/programmers to work with them interactively. To solve this problem, Eetvelde and Janssens (2004) presented a method to restructure graphs hierarchically, making them more usable and readable for the user. In Gorp, Stenten et al. (2003) they defined a way to specify refactorings of UML models. To do so, an ad-hoc extension of the UML metamodel was performed to link the source code of an application with the UML diagrams that represent it. This extension became unnecessary after UML version 1.5 because the addition of Action Semantics allowed such links to be represented. Refactorings are defined by operations composed of a precondition, a postcondition and a bad smell condition, which is specified as an OCL operation that returns a value according to a certain metric. This value determines how close a given element (the one used as a parameter of the operation) is to being refactored. With this information, the system can find and execute certain refactoring operations semi-automatically, using the following algorithm:

1. The user specifies a threshold value for each source code smell of the refactoring operations
2. A tool evaluates all source code smell operations for all parameters.
3. A refactoring may be executed on a given set of model elements when the value of the source code smell operation is greater than the threshold value and its precondition is true.
4. After each refactoring operation is executed, its postcondition is evaluated to determine whether the refactoring operation has been executed successfully and maintains program behavior.

This approach can detect some refactoring opportunities for conceptual schemas and their associated source code. However, it is not clear whether all bad smells can efficiently be represented as OCL operations.

Integrity Constraints Refactoring

One relevant part of a conceptual schema is its integrity constraints. The integrity constraints define the conditions that the instances of the schema have to satisfy in order to be correct. Integrity constraints have been the most forgotten element in the refactoring of conceptual schemas. Correa and Werner are the first who proposed refactor integrity constraints (Correa and Werner, 2004). In particular, their work deals with integrity constraints written in the OCL language. The authors adapted a few classical software refactoring operations (Fowler, 1999) to OCL and presented a set of OCL smells for these operations. The OCL smells are equivalent to Fowler's code smells: rules that imply that some OCL constraint can probably be improved through refactoring.

The OCL smells presented in this paper are:

1. *Magic literal*. It detects when an integrity constraint uses a literal within the constraint source.
2. *And chain*. It detects when a constraint consists of two or more subconstraints linked by the operator *and*.
3. *Long journey*. It detects when an OCL expression traverses a large number of associations.
4. *Rules exposure*. It detects when the business rules are specified in the postconditions or preconditions of the system operations.
5. *Duplicated code*. It detects when OCL expressions are duplicated throughout the conceptual schema.

Correa and Werner (2004) also defined five operations that deal with the abovementioned OCL smells:

- *Add variable from expression*: It adds a variable declaration to an OCL expression.
- *Replace expression by variable*: It replaces part of an OCL expression with a variable that explains its content.
- *Split AND expression*: It splits an integrity constraint made up of two or more constraints connected by *ands*.
- *Add operation definition from expression* and *Replace expression by operation call expression*: These operations deal with operations and have the same function as the first and second operations listed above.

As mentioned above, this paper only adapts part of software refactoring to OCL in order to improve the quality of OCL constraints. The list of OCL smells and operations is incomplete because there are other software refactoring operations whose adaptation to OCL might be very useful, such as the operations *Substitute algorithm* and *Remove double negative*. The former replaces a piece of the source code with another code with the same meaning; the latter eliminates double negatives used in a source fragment, thereby greatly improving its readability.

Refactoring and Quality

Software refactoring operations restructure programs in order to improve their quality. Software quality is relative and depends on the point of view from which the program is examined. For example, for one person a quality program is one that performs its tasks as quickly as possible, but for another a quality program is easy to understand and modify.

Refactoring operations do not affect all quality factors equally. A refactoring operation that improves the sharing and readability of code may worsen time performance; or an operation that improves execution time may worsen readability. In consequence, there is significant dependency between the quality factors we want to improve, the refactoring operations we need to apply and the operations we need to avoid, in order to improve the desired quality factors.

The relationship between refactoring operations and quality factors was not studied until refactoring had reached a certain maturity. Nevertheless, today it is one of the most prominent refactoring topics. The next paragraphs discuss the major works that deal with the relationship between the quality of software and refactoring operations. Other works about refactoring and quality not addressed herein are (Tahvildari, 2003; Zimmer, 2003).

Simon et al. (2001) were among the first researchers to take quality metrics into account in the refactoring process. They defined a framework that uses object-oriented metrics to detect certain refactoring opportunities for the operations *Move attribute*, *Move method*, *Extract class* and *Inline class*. In particular, they use distance-based cohesion metrics to determine when refactoring may be applied.

Du Bois (2004) proposed a technique for rewriting the postconditions of refactoring operations in order to infer automatically how the execution of a refactoring operation will affect software quality in terms of the desired metrics. Bois, Demeyer et al. (2004) adapted this technique to create a set of refactoring policies to take into account in the process of detecting refactoring opportunities. These policies were created to improve the cohesion and coupling of the system to be refactored. Du Bois' idea is promising, but it deals with few refactoring operations (five in particular) and only defines six policies that do not cover all cases.

Other works focus on how to detect refactoring opportunities and execute their associated operations taking into account the non-functional requirements of the system being refactored. In this context, Yu, Mylopoulos et al. (2003) presented a framework in which the designer used soft-goal interdependence graphs (SIG) (Chung, Nixon et al., 2000) to identify the refactoring operations that could be applied to a program in order to satisfy the non-functional requirements expressed in the SIG. SIGs are graphs that show the dependencies between goals, subgoals, resources, etc. The authors also presented a four-step algorithm for identifying and executing refactoring operations. This technique is totally manual. Moreover, it is difficult to apply due to two problems: 1) it is not clear which refactoring operations can improve a given aspect of the software, and 2) it is not clear how many operations are necessary to satisfy a given soft goal. Moreover, in the general case, the convergence of the process is not guaranteed because there may be operations that improve and worsen inverse concepts. In such cases, the application of a refactoring operation may satisfy a soft goal g_1 but violate another one g_2 and the application of another refactoring operation to satisfy the soft goal g_2 may worsen and therefore violate g_1 (which had previously been satisfied). In such a case, the algorithm will not converge.

Refactoring in Other Contexts

After software refactoring has become mature, refactoring has been applied to other contexts, such as requirements gathering (Yu, Li et al., 2004), databases, product lines (Critchlow, Dodd et al., 2003), aspects (Deursen, Marin et al., 2003) and even improving the binary code of executable programs to increase their performance.

Due to the close relationship between database refactoring and ontology refactoring we give special attention to the refactoring work in such a context. One of the first problems with database refactoring is the high degree of coupling that databases have with other sources, such as the source code of its information system, the source code of other information systems, the source code of the processes of loading and extracting data, backup processes, persistence layers, database schemas, scripts for migrating data, documentation and so on. This high degree of coupling makes it very difficult to:

1. Quantify the extent to which the application of a refactoring operation improves the quality of a database and the applications that use it, and
2. Apply refactoring, because it may be too expensive to identify, modify and rebuild all of the sources that use the refactored elements.

Furthermore, as Fowler said, refactoring a database implies restructuring not only the database schema but also the code of the applications that use the database (Fowler and Sadalage, 2003).

Fowler and the company Pramod Sadalage carried out a preliminary study on database refactoring. The study examined the application of agile techniques in the creation of the database of an information system at Pramod. The study concluded that database refactoring affects more people than software refactoring and therefore good communication is required between all stakeholders involved in the project (programmers, analysts, database administrators, etc.) in order to refactor the database successfully. This time, the definition that Fowler presents of database refactoring is too ambiguous and informal. In particular, they defined a database refactoring operation as any operation that can be executed on a database. This definition includes additive, subtractive and neutral operations. This seems to contradict our conceptual view of database refactoring: a modification of a database that maintains its behavior (the intension of the database) and the semantics of the data (the extension of the database). Obviously, Fowler's definition of database refactoring does not satisfy this statement because not all additive and subtractive operations maintain the behavior and data semantics of a database.

Ambler studied database refactoring in greater detail (Ambler, 2003) and defined database refactoring more coherently[8]:

Refactoring *is the process of changing the database schema that improves its design while retaining both its behavioral and informational semantics.*

Ambler also defined a catalog of database refactoring operations divided into five categories:

1. *Structural refactorings*: Operations that modify the database structure.
2. *Architectural refactorings*: Operations that modify the database architecture.
3. *Data quality refactorings*: Operations that improve the quality of the data stored in the database. This improvement is made using codes, standard types or other means.
4. *Referential integrity refactorings*: Operations that modify how the referential integrity constraints of the database are validated.

5. *Performance refactorings*: Operations that improve the performance of the database.

Most of Ambler's refactoring operations are adaptations of the software operations, but he presents a very interesting set of new operations. However, we believe that some operations of the catalog should be revised because it is not clear whether they are information-preserving, as they may imply a change in the functionality or semantics of the data stored in the database. For example, the operation *Remove application specific constraint* deletes integrity constraints and is applied when a business rule associated with an integrity constraint contradicts any of the other applications that share the database.

DISCUSSION

In this section we have seen the most prominent research in refactoring operations during last years. From program restructuring to database refactoring passing thru schema evolution operations, refactoring operations and conceptual schema operations. As aforesaid, most of the operations presented in these technologies are the same but with different names and few variations. Even though this disparity of names, the creation of a software refactoring catalog has establishes a kind of de facto standard naming of operations that posterior refactoring techniques such as database and conceptual schema refactoring tried to follow. The only difference we can find in the operations of each of the presented technologies are the kind of artifact to be refactored and whether the operation is bidirectional or unidirectional. For example schema transformation operations are bidirectional and therefore a schema transformation operation can transform schema S1 to another schema S2 and undo the change by transforming schema S2 to schema S1. Refactoring operations, however, are unidirectional. Therefore, if an operation transforms schema S1 to S2 (*inline class* for example), then in order to transform schema S2 to S1 we need to execute the inverse operation, which is the operation (*extract class*). We also have seen that the big challenge in refactoring is not to execute the refactoring operations but to identify their opportunities, that is what refactorings to apply and where.

Even though there have been some attempts to use refactoring in the context of ontologies (Ostrowski., 2008; Ondrej et al, 2008; Conesa, 2004), its application to ontologies is quite immature. The main problems we find in the application of the refactoring to onotologies are two: 1) it is not defined what refactoring means in the contexts of ontologies, and 2) it lacks of a catalog of refactoring operations similar to the ones create for software or databases. For example, in Protegé we can find several functionalities for refactor ontologies. For example, version 4 of Protégé allows execute 13 refactoring operations. Even though some of them can be seen as an adaptation of software refactoring the name conventions they have used are quite different, making very difficult to understand what is the exact meaning of a given refactoring operation even when the user has experience in the refactoring field.

Table 2 shows a brief comparison between the techniques presented in this section and shows, for each technique, the artifact to be refactored, whether there exists an agreed definition of its purpose, whether the operations are bidirectional and if there exists an agreed catalog of operations.

In conclusion, in order to easily apply refactoring to ontologies, a catalog that contains the refactoring operations applicable to conceptual schemas must be created. This catalog should be similar to the catalog of software refactoring[9]. Such a catalog should identify for each refactoring operation its preconditions (when refactoring can be applied to a given set of ontology concepts), the steps for executing the refactoring on ontologies (its effect) and its expected result (postconditions). Furthermore, we think a deeper study of conceptual schema refactoring is necessary to

Table 2. Summary of the main refactoring techniques dealt in the paper

	Refactored artifact	Agreed definition	Directionality of application	Existence of a catalog of operations
Program restructuring	Programs	Yes	Unidirectional	
Schema evolution	Conceptual schemas (usually database and OO schemas)	Yes	Bidirectional	Several small catalogs of operations
Software refactoring	Programs	Yes	Unidirectional	Yes
Conceptual schema refactoring	Conceptual Schemas (usually UML schemas)	No	Unidirectional	No
Database refactoring	Databases and related programs	Yes	Unidirectional	Yes
Ontology refactoring	Ontologies	No	Unidirectional	No

detect specific new smells of conceptual schema refactoring.

ONTOLOGY REFACTORING

Up to now, refactoring has been applied to a wide range of sources and artifacts: software, UML models, use cases, etc. Wikipedia[10] offers the refactoring definition that we feel best identifies its meaning regardless of the artifact to be refactored:

Refactoring *is the process of rewriting written material to improve its readability or structure, with the explicit purpose of keeping its meaning or behavior.*

All artifacts that have been refactored up to now follow a grammar and can therefore be expressed as a textual representation, making the above definition applicable in all cases. However, each artifact tends to have its own definition of refactoring that clearly specifies the meaning of *meaning* and *behavior* in its context. For example, software refactoring is defined as *the process of restructuring software to make it easier to understand and cheaper to modify without changing its observable behavior.* Since the purpose of all software systems is to offer a predetermined behavior – that is, always returning the same outputs for a given set of inputs – the correspondence between the two refactoring definitions is clear. In fact, this definition is a specialization of Wikipedia's definition, because it has the same meaning but clearly specifies the quality factors that a software refactoring operation should improve and the definition of *meaning* and *behavior* of the software.

In the database field, Ambler (2003) redefined refactoring as *the process of changing the database schema that improves its design while retaining both its behavioral and informational semantics.* This definition is not very precise, but like the previous one, it specifies the quality factors to be improved (the design) and the meaning and behavior of the database, defined as the informational semantics (the same information base is expressed) and the behavioral semantics (triggers, stored procedures, etc.).

Ontologies are neither programs nor databases. Therefore, their refactoring definition should be refined, as in the aforementioned cases. The abstraction level of an ontology is higher than that of a database schema or a program. Therefore, all quality factors related to a particular implementation of a conceptualization or the representation of a domain population are irrelevant to the quality of an ontology, because its goals are essentially semantic. Hence, we can define ontology refactoring as: *the process of restructuring ontologies to improve their readability or structure, while preserving their relevant knowledge.*

Ontologies are designed to satisfy a set of requirements. A change to an ontology preserves its relevant knowledge if the knowledge necessary to satisfy its requirements is not deleted in the process. In other words, a refactored ontology should be capable of inferring the same knowledge as the original ontology. Therefore, the purpose of ontology refactoring is not to change the associated conceptualization (*conceptual change*) (Thagard.P, 1992) but rather the way in which the associated conceptualization is represented (*representational change*).

One of the most difficult tasks in the execution of refactoring operations is modifying the source code that refers to the refactored elements. For example, if we execute the refactoring operation *Change method* (Fowler, 1999) to change the name of the method *GetPersons()* to *GetPeople()*, all references to the method *GetPersons()* in the source code must be identified and replaced with references to the method *GetPeople()*. This problem also occurs in ontology refactoring, even when the ontology does not define behavioral information, because ontologies may contain general integrity constraints for example some SWRL rules. These constraints may refer to any concept of the ontology. Therefore, after a concept is changed, all integrity constraints that refer to that concept must be modified to maintain the syntactic consistency of the ontology.

Another problem in ontology refactoring is the fact that tests cannot be performed to check that the ontology's behavior has not been modified. Tests of this kind are necessary in software and database refactoring, but cannot be applied to ontologies because, as mentioned above, ontologies cannot be executed. However, a validation of some sort may be carried out on various ontology instantiations to check, after each execution of a refactoring operation, whether any instances have been lost as a consequence of the change. To perform this test, the refactoring operation must not change only the structure of the ontology but also its information base (instances of the ontology).

A Catalog of Ontology Refactoring Operations

As mentioned above, some works have dealt with the application of refactoring operations to conceptual schemas and ontologies. However, an exhaustive catalog of refactoring operations applicable to ontologies has not yet been defined.

This section presents a catalog of refactoring operations applicable to ontologies adapted from the fields of software refactoring, database refactoring, OCL refactoring, schema transformation and program restructuring. This catalog presents the operations defined in these areas that can improve certain aspects of an ontology or conceptual schema while maintaining its semantics. In the particular case of software refactoring, some operations have been rejected because they focus on implementation and therefore do not make sense on the abstract level of an ontology.

At first glance, the proposed operations may seem too simple and specialized to cause any improvement. However, as Fowler (1999) mentioned in his book, refactoring steps tend to be small, but they can be combined in sequences to construct advanced refactoring operations that can satisfy more ambitious goals and use higher-level tactics.

Since Fowler's classification of software refactoring operations has become the standard de facto in software refactoring, we used the categories of Fowler's classification in the catalog of ontology refactoring operations with classification purpose. We have extended fowler classification by adding a new level in order to deal with the refactoring operations that deal with general integrity constraints.

In particular, we group ontology refactoring operations in two main categories:

- **Structural refactoring operations:** These operations can refactor the structural ele-

Figure 1. presents the taxonomy we use to classify refactoring operations

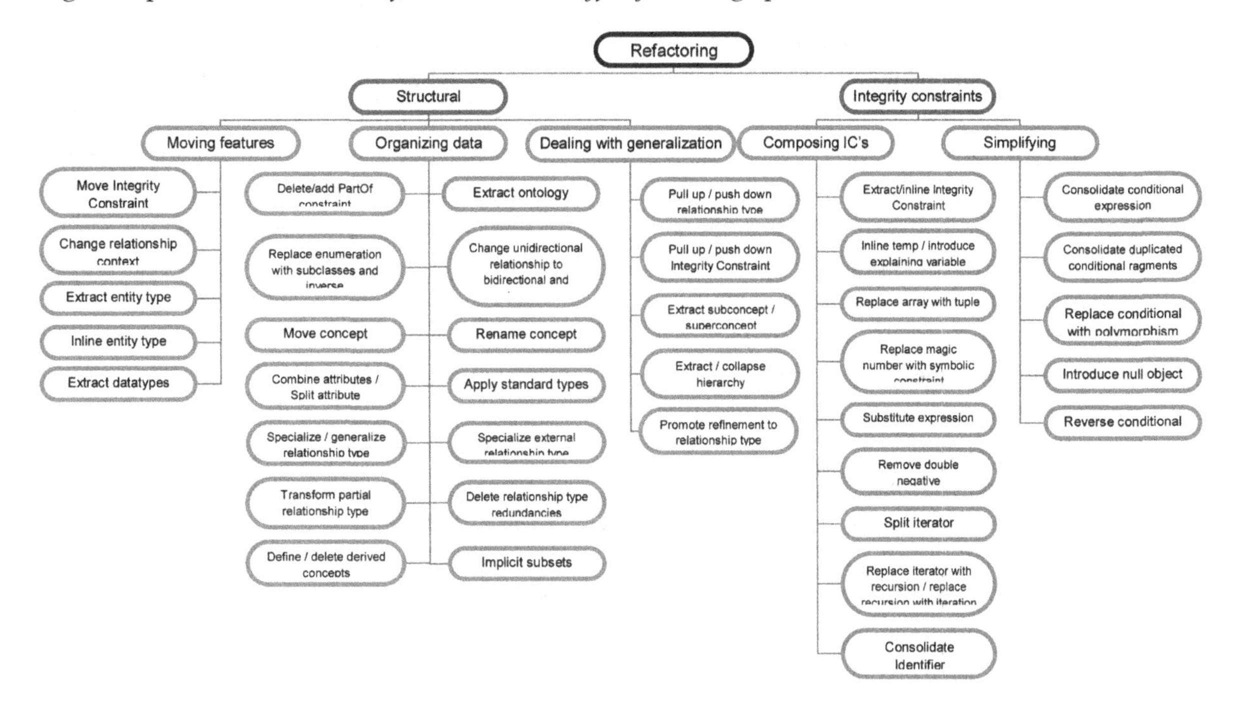

ments of an ontology (concepts, relationship types, generalization relationships and instances and constraints embedded in the ontology language).

- **Integrity constraint refactoring operations:** These operations can refactor the dynamic elements of an ontology such as derivation rules written in SWRL or OCL and its general integrity constraints.

The first group of refactoring operations can be applied to any ontology, and the second group is only relevant in ontologies that contain general integrity constraints or operations. However, we have not considered the refactoring of operations because it has little practical application in today's ontologies.

We have kept certain refactoring operations even though the elements they deal with are theoretically not very relevant to ontologies because they belong to a lower abstraction level. Examples of such elements include directed relationship types and iterations. These operations have been kept because such elements exist in some current ontologies.

The catalog of software refactoring is currently the largest and best-documented refactoring catalog even created. Therefore, most of the refactoring operations in our catalog have been adapted from it. This catalog is the origin of all refactoring operations unless otherwise specified. When a single operation has been defined for more than one artifact, we use the definition that is closest to the operation's meaning in the context of ontologies.

Structural Refactoring Operations

Structural refactoring operations are those which modify the structural concepts of the ontology – that is, its classes, its relationship types, its instances, and the integrity constraints that are embedded in the ontology language, such as the inheritance relationships that determine that a concept is a subtype of another concept, or the cardinality constraints of a relationship type.

The structural relationship refactoring operations are classified within three main categories according to their purpose:

1. Moving features between concepts: operations that change the context in which the properties of an ontology (relationship types and integrity constraints) are defined.
2. Organizing data: operations that change the ontology structure in order to make it easier to work with its data.
3. Dealing with generalization: operations used to change the taxonomy or move ontology properties (relationship types and integrity constraints) through the taxonomy of the ontology. Note that both classes and relationship types may have taxonomical structures in ontologies.

The operations of each category are explained in the following lines.

Moving Features between Concepts

- *Move Integrity Constraint*: This operation moves an integrity constraint from one entity type to another. This is an adaptation of Fowler's *Move method* operation.
- *Change relationship context*: This operation changes the type of one of the participants in a given relationship type. The old and new participants in the relationship type cannot be related with a generalization/specialization relationship. This operation is based on Fowler's *Move field* operation. In our approach, we see an attribute or field as a special case of a binary relationship type.
- *Extract entity type*: This operation splits an entity type in two and distributes its content between the two new entity types. It is based on Fowler's *Extract class* operation.
- *Inline entity type*: This is the inverse of the operation *Extract entity type*. It merges two entity types into one new entity type. It is based on Fowler's *Inline class* operation.
- *Extract datatype*: This operation changes an entity type into a datatype. Although datatypes are a particular case of entity types, we created this operation because they have a prominent role in ontologies. In the ontologies that incorporate the concept of datatypes, such as in OWL, this operation converts a class in a datatype and changes all the relationship types that dealt with the class to attributes, or data properties in the case of OWL.

Organizing Data

- *Delete PartOf constraint*: This operation deletes a *PartOf* integrity constraint related to a given binary relationship type. In the particular case of UML ontologies, this operation replaces an aggregation with an association. It is based on Fowler's *Change value to reference* operation.
- *Add PartOf constraint*: This is the inverse of the operation *Delete PartOf constraint*. It is based on Fowler's *Change reference to value* operation.
- *Change unidirectional relationship type to bidirectional*: This operation changes the navigability of a relationship type from unidirectional to bidirectional. In the particular case of ontology languages that do not support bidirectional relationship types, like OWL, this operation creates a new relationship type between the same participants but with inverse direction and define it as the inverse of the original relationship type. It is based on Fowler's *Change unidirectional relationship type to bidirectional* operation.
- *Change bidirectional association to unidirectional*: it is the inverse of the operation *Change unidirectional association to bidi-*

rectional and based on Fowler's *Change bidirectional relationship type to unidirectional* operation.

- *Replace enumeration with subclasses*: This operation replaces an enumeration attribute used to determine the type of the instances of a given entity type *E* with a set of subentity types of *E*. Each of the entity types created represents one of the possible values of the deleted attribute. This operation was adapted from the software refactoring operation called *replace type code with subclasses*.
- *Replace subclass with an enumerated type*: This is the inverse of the previous operation. This operation replaces a set of subtypes of a common entity type *E* with a relationship type in *E* that represents the same information. An integrity constraint is created to restrict the number of possible values of the new relationship type; this number is equal to the number of deleted subtypes. This operation was adapted from the software refactoring operation *Replace subclass with fields*.
- *Extract ontology*: Some ontology languages allow an ontology's knowledge to be grouped in small blocks based on its meaning, context or domain. For example, the OWL languages allow for defining small ontologies that can be reused in other context. Therefore, this operation splits a given ontology into two subontologies.
- *Move concept*: This operation moves a concept that belongs to one ontology to another ontology.
- *Combine attributes:* This operation merges several attributes into one. It was adapted from the database refactoring operation *Combine columns representing a single concept*. This operation is very useful when a designer creates relationship types too specific for the objectives of the ontology. For example, a designer may define phone numbers using three relationship types: one to identify the country, another to identify the city and finally the number without these two prefixes. Before the refactoring operation, three relationship type instances were needed to identify a phone number. For example, a phone number in Atlanta (USA) would be represented by 1, 404 and 1234567. After the refactoring operation, this phone number would be represented using only one relationship: 1404123456.
- *Split attribute*: This is the inverse of the operation *Combine attributes*. It splits 1 attribute into *n*, with *n*>1. The types of the new attributes should be the same as the original or any of its subtypes. This operation is based on the database refactoring operation *Split column*.
- *Rename concept*: This operation changes the name of a concept (an entity or relationship type) and it has been adapted from the database refactoring operations *Rename attribute* and *Rename table* (Ambler, 2003).
- *Apply standard types*: Sometimes, different attributes that represent the same (or similar) concepts have different types. For example, a phone number may be represented by a *String* and a cell phone number by an *Integer*. Obviously, the quality of the ontology is improved if represent all phone numbers in the same way. This refactoring operation solves this problem, changing the type of different attributes that represent similar things until all of them have the same types or share a common supertype. In this example, this refactoring operation would change the end of one relationship type from *Integer* to *String* and therefore all phone numbers of the ontology would be represented in the same way. This operation has been adapted from database refactoring.
- *Relationship type specialization*: This operation change one *n*-ary relationship type for *m* (*n*-1)-ary more specific relationship types. This operation is particularly useful

when one participant in a relationship type can only have a predefined set of values known prior to the conceptualization phase. In such a case, the *n*-ary relationship may be replaced with *m* relationship types, where *m* is the number of possible values of the participant. Figure 2 shows a fragment of an ontology that represents the number and kinds of medals each country has won in the Olympic Games. The entity type *Medal Kind* can only have three instances: *Gold, Silver* and *Bronze*. Hence, we can apply this refactoring operation to replace the ternary relationship type *Won* with three binary relationship types: one for each possible instance of the *Medal kind* entity type. The new relationship types are called *Won gold in, Won silver in* and *Won bronze in* and represent the number of medals each country won of a each type. This operation was adapted from Halpin's schema transformation operations (Halpin, 2001).

- *Relationship type generalization*: This is the inverse of the previous operation (*Predicate specialization*). It replaces *m n*-ary relationship types with similar semantics with one relationship type with an arity of *n+1*. The new participant in the relationship type is used to identify the semantics (previous relationship type) used in each instance of the relationship type. This operation was adapted from Halpin's operation *Predicate specialization* (Halpin, 2001).
- *External relationship specialization*: This operation makes more specific a given relationship type based on the values of another relationship type. To do this, the first relationship type absorbs the second one. This operation was adapted from the schema transformation field (Halpin, 2001).
- *Transforming partial relationship type*: This operation changes a partial relationship type (with a minimum cardinality of 0) into a total one (with a minimum cardinality of 1).

Figure 2. Example of the application of the refactoring operation Relationship type specialization

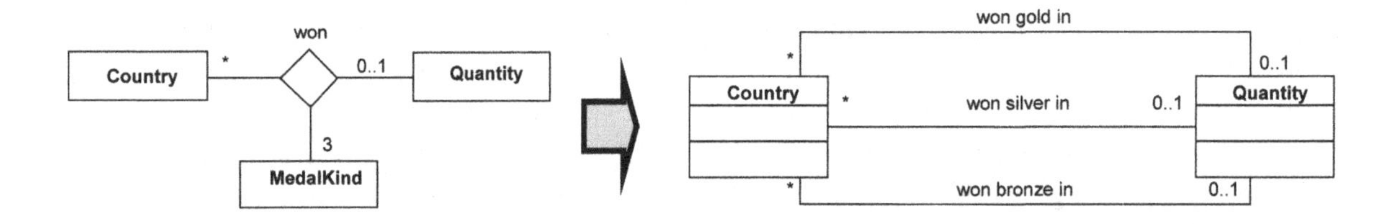

Figure 3. Example of Transforming Partial Relationship Type refactoring operation

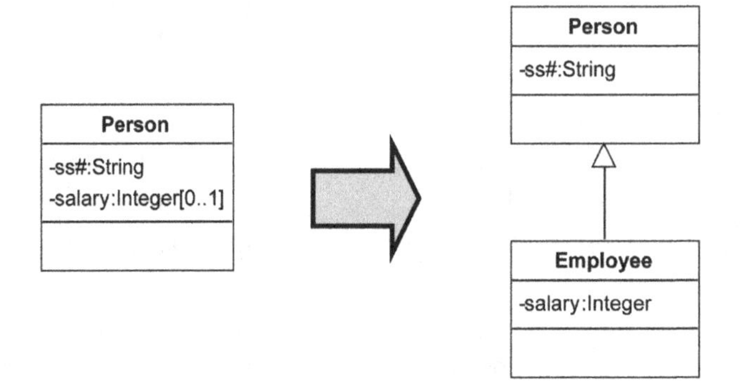

To do so, a new subtype that contains the instances that used to participate in the partial relationship type is created. Thereafter, the relationship type is redefined by pointing the new created subtype and changing the minimal cardinality to one. Figure 3 shows an example of application of this refactoring operation to the partial attribute *salary*. This operation was adapted from the schema transformation area, specifically from the operation *Transforming partial attributes* defined in Assenova and Johannesson (1996).

- *Transforming partial relationship types that are total in union*: Sometimes, an integrity constraint can represent that the union of two partial relationship types is total, which means that for each instance of a partial participant there is at least one instance of these relationship types that relates it. In this case, this operation makes sense and replaces the two partial relationship types with one total relationship type. It also creates a new entity type defined as the union of the entity types that participated in the previous relationship types. This refactoring operation also deletes the integrity constraint that indicated the totality of the union of the two relationship types. For example, the relationship types *ResponsibleGC* and *ResponsibleUC* shown in Figure 4 are partial, but the integrity constraint defined in the figure states that their union is total because a *Head teacher* is responsible for a course. By applying this operation the entity type *Course* is created as the union of *UndergraduateCourse* and *GraduateCourse,* and the two partial relationship types are replaced with the total

Figure 4. Example of Transforming partial relationship types that are total in union refactoring operation

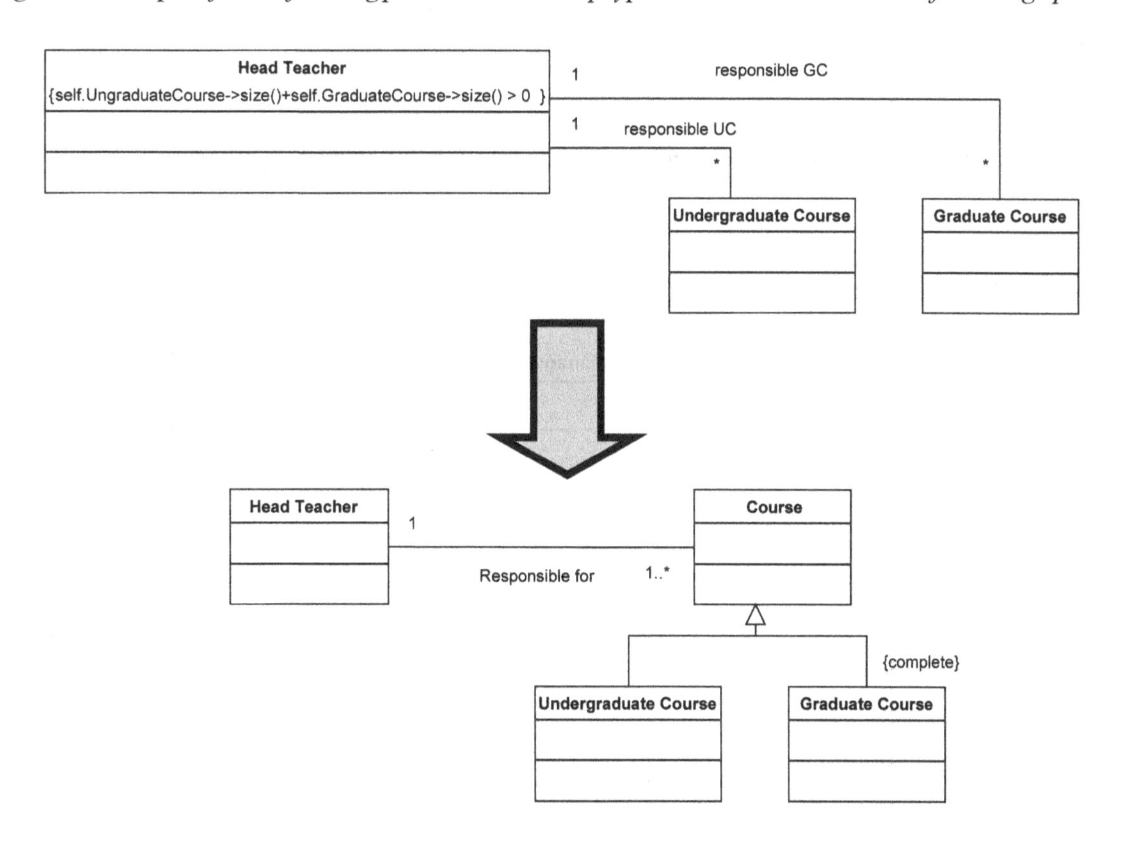

relationship types *ResponsibleFor*. This operation was adapted from Assenova and Johannesson's list of schema transformation operations (Assenova and Johannesson, 1996).

- *Deleting relationship types redundancy*: This operation, based on a schema evolution (Batini, Ceri et al., 1992), deletes redundant relationship types. To be redundant, two relationship types must share both the same participants and semantics. Therefore, this operation always requires designer intervention to determine whether the *n* relationship types, which are supposedly redundant, have the same semantics. For example, there are two paths between the entity types *Employee* and *Department* of Figure 5 with the following meanings: 1) the department where the employee works (*WorksIn*), and 2) the department the employee's manager directs (*DirectedBy->Heads*). If the employee's manager works in the same department as all of his or her subordinates, then the two paths represent the same information and are therefore redundant. In such a case, we can apply this refactoring operation to delete the redundant relationship type *WorksIn*.
- *Defining derived concepts*: This operation defines a new entity type or relationship type whose population is derived from the information base of the ontology. It can be applied, for example, to the relationship type *age* of an entity type *Animal* in order to state that the age of an animal is a derived relationship type obtained by subtracting the animal's birth date from today's date. This operation is based on the schema transformation operation *defining derived concepts* (Batini, Ceri et al., 1992).
- *Deleting derived concepts*: This operation deletes a derived concept (entity type or relationship type). This is possible when derived concept is redundant in a conceptual schema. If it is not relevant, then its elimination does not imply any loss of ontology semantics.
- *Implicit subsets*: This operation deletes redundant generalization paths. It was adapted

Figure 5. Example of refactoring operation for deleting relationship types redundancy

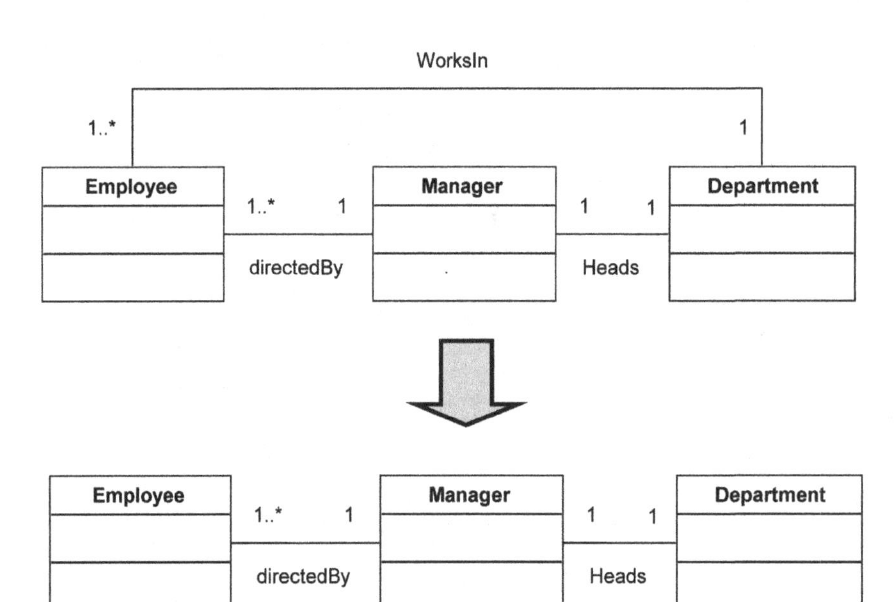

from the schema transformation operations of Batini, Ceri et al. (1992).

Dealing with Generalization

- *Pull up relationship type*: Conceptually speaking, this operation replaces one participant in a relationship type with its supertype. This operation is based on the software refactoring operation *Pull up property end*, but it needs to be modified completely in order to adapt it to ontologies and take into account the possible integrity constraints that can affect it: disjointness and completeness of classes.
- *Push down relationship type*: This operation replaces one participant in a relationship type with the subtypes of that participant. It is the inverse of the operation *Generalize relationship* and is based on the software refactoring operation *Push down property end*.
- *Pull up IC*: Conceptually speaking, this operation replaces the context of a relationship type with the supertype of that context.
- *Push down IC*: This operation replaces the context of a relationship type with the subtype of that context. This operation, which we created, is the inverse of the integrity constraint *Pull up IC*.
- *Extract subconcept*: This operation creates a new concept as a subtype of an existing one and moves some of the existing concept's properties to the new one. This operation is based on the software refactoring operation of the same name.
- *Extract superconcept*: The operation creates a new concept as a supertype of an existing one and moves some of the existing concept's properties to the new one. It is the inverse of the operation *Extract subconcept* and has its origins in the software refactoring operation *Extract superconcept*.
- *Collapse hierarchy*: This operation collapses one concept (entity type or relationship type) with its subtypes. The new concept contains the semantics of the two collapsed elements. There are two versions of this operation: *Collapse hierarchy up* and *Collapse hierarchy down*. In the "up" version, the supertype absorbs the subtypes and therefore the new concept takes the name of the supertype. In the "down" version is the subtype the one that keeps its name. This operation is based on the software refactoring operation *Collapse hierarchy*.
- *Extract hierarchy*: This operation splits a concept into several concepts related by generalization relationships and spreads the properties of the original concept to the new concepts. It is the inverse of the operation *Collapse hierarchy* and has its origin in the software refactoring operation of the same name.
- *Promote refinement to relationship type*: The relationship types defined in an ontology tend to be used, with some restrictions, by the subtypes of the entity types where they are defined. In such cases, a relationship type may be redefined using either a predefined construction of the ontology or a general integrity constraint. When a relationship type is only used by the elements where it is redefined or their subtypes, then it and its redefinition can be replaced with another relationship type that uses as participants the entity types where the relationship type was redefined and takes into account the integrity constraints added in the redefinition. We created this operation from scratch because it is highly applicable in large and general ontologies.

Integrity Constraint Refactoring Operations

The operations in this category can be used to restructure the contents of general integrity constraints. We call general integrity constraints to integrity constraints which cannot be represented directly with the ontology language and needs to use a new language to express them, such as the OCL language in the case of UML ontologies. These operations are useless when the ontology being refactored does not allow general integrity constraints, as in the case of an ontology that only uses OWL language.

The integrity constraint refactoring operations are classified within two main categories according to their purpose:

1. Composing integrity constraints: operations that improve the quality of integrity constraints.
2. Simplifying integrity constraints: operations that simplify the conditions and expressions used within integrity constraints.

The operations presented in this section are generic enough refactor general integrity constraints written in either imperative or declarative languages. Next two subsections enumerate and describe briefly the refactoring operations.

Composing Integrity Constraints

- *Extract IC*: This operation splits one integrity constraint in two and spreads its code to the two created constraints. This operation is based on the characteristics of the software operation *Extract method.*
- *Inline IC*: This operation merges two integrity constraints into one. It is the inverse of the operation *Extract IC.* It is based in the software refactoring operation *Inline method.*
- *Inline temp*: This operation replaces a temporal variable used in an integrity constraint with the expression that defines its value. This operation is very useful when the temporal variable is used infrequently. It was adapted from the software refactoring operation of the same name.
- *Introduce explaining variable*: This operation assigns a variable the result of an expression used in an integrity constraint and replaces the expression with a reference to the new variable wherever it occurs. This operation improves maintainability when the same expression is used several times in an integrity constraint. It is the inverse of the operation *Inline temp* and is based on the software refactoring operation *Introduce explaining variable*.
- *Replace array with tuple*: This operation replaces an array used in the body of an integrity constraint with a tuple. The new tuple will have one field for each row of the array. We included this operation in our catalog because some of the languages for representing general integrity constraints may support tuples and arrays. It is an adaptation of the software operation *Replace array with object*.
- *Replace magic number with symbolic constant*: This operation replaces a number used in an integrity constraint with a constant of the same value. It was adapted from the software operation of the same name.
- *Substitute expression*: This operation replaces an expression of an integrity constraint with another expression with the same meaning. It is an adaptation of the software refactoring operation *Substitute algorithm.*
- *Remove double negative*: Using the *DeMorgan* rule, this operation replaces a double negation with an affirmation, for example replacing *if(Not NotFound)* for *if(Found)*. This operation was adapted from the software refactoring operation of the same name.
- *Split iterator*: This operation separates one iteration in n (n>1), where each of the n

new iterators performs a different activity. We decided to include in the catalog some operations that deal with iterators because they are often used in the languages that represent general integrity constraints. This operation is based on the software refactoring operation *Split Loop*.

- *Replace iteration with recursion*: This operation replaces an iteration structure with its equivalent recursive call. It was adapted from the software refactoring operation of the same name.
- *Replace recursion with iteration*: This operation replaces a recursive call structure with its equivalent iteration structure. It is the inverse of the operation *Replace iteration with recursion*.
- *Consolidate identifier*: This operation can be applied when two or more expressions use different identifiers to refer to the same concept. It forces all expressions to access the instances of a concept in the same way, which means using the same identifier. This operation is based on the database refactoring operation *Consolidate key strategy* (Ambler, 2003).

Simplifying Conditional Expressions

- *Consolidate conditional expression*: This operation combines *n* conditionals defined within an integrity constraint in a single conditional. It was adapted from the software refactoring operation of the same name.
- *Consolidate duplicate conditional fragments*: This operation extracts the parts of a conditional structure that are repeated in all of its branches. It was adapted from the software refactoring operation of the same name.
- *Replace conditional with polymorphism*: In some cases, a conditional states restrictions that depend on the type of the individual. This operation distributes these restrictions through the taxonomy using the inheritance as the template design pattern (Gamma, Helm et al., 1995). In particular, this operation distributes the conditional through the subtypes of the context entity type of the integrity constraint where it is defined. Thus, for each subtype, the conditions that its instances should satisfy are defined. This operation was adapted from the software refactoring operation of the same name.
- *Introduce null object*: This operation creates a new subtype of a given entity type *E*. This subtype represents the instances of *E* with undefined information, that is, the instances that have no value for a given relationship type. This operation is very useful for deleting conditional constraints such as "*if (something=null) then...*" It was adapted from the software refactoring operation of the same name.
- *Reverse conditional*: This operation modifies a conditional to make it more comprehensible. It negates the entire conditional and therefore the condition and the *then* and *else* branches.

CONCLUSION

Over the last two decades refactoring operations have been applied effectively to software. Refactoring has also been applied successfully, but in a minor extent, to other artifacts such as databases and UML models. There have been some attempts to apply refactoring to ontologies, but with little success. The main problems in applying refactoring to ontologies are that it is not clear what refactor an ontology means, and the necessity of a catalog of refactoring operations similar to the ones created for software or databases.

In this chapter we studied the refactoring work of other fields in combination with other techniques such as program restructuring or schema transformation. The lessons learnt of this study have been used to define what ontology refactoring is and to present a catalog of ontology refactoring operations, which reuse the operations presented in the fields of software refactoring, database refactoring, model refactoring and schema transformation that are suitable to refactor ontologies. Therefore, the main contributions of this chapter are clarification of the meaning of ontology refactoring and the creation of a catalog that contains ontology refactoring operations. The catalog has been created following the classification and naming conventions of previous refactoring catalogs to make the new catalog easier to use and to understand.

The presented work provides a first step in the formalization of ontology refactoring, but more work needs to be done in that direction. In particular, we believe that the coherent next step would be to define formally the operations of the presented catalog and implement them in the most prominent ontology tools, such as Protégé. In addition, the identification of ontology refactoring opportunities needs also to be addressed in order to identify in what cases some refactoring operations can be applied to automatically improve ontologies.

ACKNOWLEDGMENT

This work has been partly supported by the Spanish Ministry MICINN (TIN2008-00444) and the European Union (FP7-ICT-2009-5-257639).

REFERENCES

Ambler, S. (2003). *Agile database techniques, effective strategies for the agile software developer*. Wiley.

Ambler, S. W. (2002). *Agile modelling, effective practices for extreme programming and the unified process*. John Wiley & Sons Inc.

Assenova, P., & Johannesson, P. (1996). *Improving quality in conceptual modelling by the use of schema transformations*. 15th International Conference on Conceptual Modeling.

Astels, D. (2002). *Refactoring with UML*. Conference on eXtreme Programming and Agile Processes in Software Engineering XP2002, (pp. 67-70).

Batini, C., & Ceri, S. (1992). *Conceptual database design, an entity-relationship approach*. Benjamin/Cummings.

Beck, K. (1999). *Extreme programming explained*. Addison-Wesley Pub Co.

Bergstein, P. L. (1991). Object-preserving class transformations. *Conference proceedings on Object-oriented programming systems, languages, and applications*. Phoenix, Arizona, United States, ACM Press, (pp. 299-313).

Bois, B. D. (2004). Opportunities and challenges in deriving metric impacts from refactoring postconditions. In S. Demeyer, S. Ducasse & K. Mens (Eds.), *Proceedings WOOR'04, ECOOP'04 Workshop on Object-Oriented Re-engineering*. Antwerpen, Belgium.

Bois, B. D., Demeyer, S., et al. (2004). Refactoring-improving coupling and cohesion of existing code. In *Proceedings WCRE'04 Working Conference on Reverse Engineering*. Victoria, Canada, IEEE Press, (pp. 144-151).

Bottoni, P., Parisi-Presicce, F., et al. (2002). Coordinated distributed diagram transformation for software evolution. In R. Heckel, T. Mens & M. Wermelinger (Eds.), Proceedings of the Workshop on 'Software Evolution Through Transformations'(SET'02), (p. 72).

Casais, E. (1989). Reorganizaing an object system. In Tsichritzis, D. (Ed.), *Object oriented development* (pp. 161–189). Geneve.

Chung, L., & Nixon, B. A. (2000). *Non-functional requirements in software engineering*. Kluwer Academic Publishers.

Conesa, J. (2004). *Ontology-driven Information Systems, pruning and refactoring of ontologies*. Doctoral Symposium of the Seventh International Conference on the Unified Modeling language. Retrieved from http.//www-ctp.di.fct.unl.pt/UML2004/DocSym/JordiConesaUML2004DocSym.pdf

Conesa, J. (2008). *Pruning and refactoring ontologies in the development of conceptual schemas of Information Systems*. Unpublished doctoral thesis. Technical University of Catalonia.

Correa, A., & Werner, C. (2004). Applying refactoring techniques to UML/OCL models. *Proceedings of International Conference of UML*. Lisbon, Portugal, (pp. 173-187).

Critchlow, M., Dodd, K., et al. (2003). Refactoring product line architectures. *Proceedings of the First International Workshop on REFactoring, Achievements, Challenges, Effects* (REFACE). Victoria, Canada.

Deursen, A. v., Marin, M., et al. (2003). Aspect mining and refactoring. *Proceedings of the First International Workshop on REFactoring, Achievements, Challenges, Effects* (REFACE03). Victoria, Canada.

Ducasse, S. e., Rieger, M., et al. (1999). A language independent approach for detecting duplicated code. In H. Yang & L. White (Eds.), *Proceedings ICSM'99 International Conference on Software Maintenance*. IEEE, (pp. 109-118).

Ducasse, S. e., Rieger, M., et al. (1999). *Tool support for refactoring duplicated OO code*. ECOOP'99, Springer. 1743, (pp. 177-178).

Eetvelde, N. V., & Janssens, D. (2004). *A hierarchical program representation for refactoring*. (ENTS 82).

Eick, C. F. (1991). *A methodology for the design and transformation of conceptual schemas*. Very Large Data Bases Conference, Barcelona.

Ernst, M. D., & Cockrell, J. (2001). Dynamically discovering likely program invariants to support program evolution. *IEEE Transactions on Software Engineering*, *27*(2), 1–25. doi:10.1109/32.908957

Fowler, M. (1999). *Refactoring, improving the design of existing code*. Addison-Wesley.

Fowler, M., & Sadalage, P. (2003). *Evolutionary database design*. Retrieved from http.//www.martinfowler.com/articles/evodb.html

Gamma, E., & Helm, R. (1995). *Design patterns, elements of reusable object-oriented software*. Addison-Wesley.

Gorp, P. V., Stenten, H., et al. (2003). Towards automating source-consistent UML refactorings. UML, 2003, Springer. 2863, (pp. 144-158).

Grant, S., & Cordy, J. R. (2003). An interactive interface for refactoring using source transformation. *Proceedings of the Fist International Workshop on Refactoring, Achievements, Challenges and Effects* (REFACE'03). Victoria, Canada, (pp. 30-33).

Gray, J. G. (2002). *Aspect-oriented domain-specific modeling, a generative approach using a meta-weaver framework. Department of Electrical Engineering and Computer Science*. Nashville: Vanderbilt University.

Griswold, W. G., & Chen, M. I. (1998). Tool support for planning the restructuring of data abstractions in large systems. *IEEE Transactions on Software Engineering*, *24*(7), 534–558. doi:10.1109/32.708568

Griswold, W. G., & Notkin, D. (1993). Automated assistance for program restructuring. *ACM Transactions on Software Engineering and Methodology*, *2*(3), 228–269. doi:10.1145/152388.152389

Halpin, T. (1996). *Conceptual schema and relational database design*. Prentice-Hall, Inc.

Halpin, T. (2001). *Information modeling and relational: From conceptual analysis to logical design*. Morgan Kaufman.

Halpin, T. A., & Proper, H. A. (1995). *Database schema transformation & optimization*. International Conference on Conceptual Modeling, LNCS.

Hofstede, A. H. M. t., Proper, H. A., et al. (2007). *A note on schema equivalence*.

Judson, S. R., Carver, D. L., et al. (2003). A metamodeling approach to model refactoring.

Kataoka, Y., & Ernst, M. D. (2001). *Automated support for program refactoring using invariants* (pp. 736–743). ICSM.

Kleppe, A., & Warmer, J. (2003). *MDA explained. The model driven architecture, practice and promise*. Addison-Wesley.

Koru, A. G., Ma, L., et al. (2003). Utilizing operational profile in refactoring large scale legacy systems. *Proceedings of the First International Workshop on REFactoring, Achievements, Challenges, Effects* (REFACE). Victoria, Canada.

Marko Boger, T. S., & Fragemann, P. (2002). *Refactoring browser for UML*. Objects, Components, Architectures, Services, and Applications for a NetworkedWorld, International Conference NetObjectDays, NODe, 2002, (pp. 77-81).

Mens, K., & Tourwe, T. (2004). *Delving source code with formal concept analysis*. Computer Languages, Systems & Structures.

Mens, T. (2004). A survey of software refactoring. *IEEE Transactions on Software Engineering, 30*(2), 126–139. doi:10.1109/TSE.2004.1265817

Moore, I. (1996). Automatic inheritance hierarchy restructuring and method refactoring. *Proceedings of the 11th ACM SIGPLAN conference on Object-oriented programming, systems, languages, and applications*. San Jose, California, United States, (pp. 235-250).

OMG. (2003). *OMG revised submission, UML 2.0 OCL*.

Opdyke, W. F. (1992). *Refactoring object-oriented frameworks*. Unpublished doctoral dissertation, University of Illinois.

Ostrowski, D. A. (2008). *Ontology refactoring*. Second IEEE International Conference on Semantic Computing (ICSC, 2008), (pp. 476-479).

Philipps, J., & Rumpe, B. (2001). *Roots of refactoring*. Tenth OOPSLA Workshop on Behavioral Semantics. Tampa Bay, Florida, USA.

Porres, I. (2002). *A toolkit for manipulating UML models*. TUCS Turku Centre for Computer Science.

Porres, I. (2003). *Model refactorings as rule-based update transformations. UML, 2003* (pp. 159–174). Springer.

Proper, H. A., & Halpin, T. A. (2004). *Conceptual schema optimisation-database optimisation before sliding down the waterfall* (p. 34). Department of Computer Science - University of Queensland.

Roberts, D., Brant, J., et al. (1997). A refactoring tool for smalltalk. *Theory and Practice of Object Systems, 3*(4).

Roberts, D. B. (1999). *Practical analysis for refactoring*. Unpublished doctoral dissertation, University of Illinois.

Rysselberghe, F. V., & Demeyer, S. (2004). Evaluating clone detection techniques from a refactoring perspective. *Proceedings of the 19th International Conference on Automated Software Engineering* (ASE'04). Linz, Austria, (pp. 336-339).

Sacco, G. (2000). Dynamic taxonomies, a model for large information bases. *IEEE Transactions on Data and Knowledge Engineering, 12*(3), 468–479. doi:10.1109/69.846296

Simmonds, J., & Mens, T. (2002). *A comparison of software refactoring tools*. Programing Tecnology Lab.

Simon, F., Steinbrukner, F., et al. (2001). Metrics based refactoring. *Proceedings of the 5th European Conference on Software Maintenance and Reengineering*, IEEE Computer Society Press.

Sunye, G., & Pollet, D. (2001). Refactoring UML models. In Gogolla, M., & Kobryn, C. (Eds.), *UML* (pp. 134–148). Springer.

Sváb-Zamazal, O., Svátek, V., Meilicke, C., & Stuckenschmidt, H. (2008). *Testing the impact of pattern-based ontology refactoring on ontology matching results*. 3rd International Workshop on Ontology Matching.

Tahvildari, L. (2003). *Quality-driven object-oriented code restructuring*. In the IEEE Software Technology and Engineering Practice (STEP) - Workshop on Software Analysis and Maintenance, Practices, Tools, Interoperability (SAM). Amsterdam, The Netherlands.

Thagard, P. (1992). *Conceptual revolutions*. Princeton University Press.

Tourwe, T., & Mens, T. (2003). Identifying refactoring opportunities using logic meta programming. *Proceedings of the 7th European Conference on Software Maintenance and Reengineering*, IEEE Computer Society Press.

Xing, Z., & Stroulia, E. (2003). Recognizing refactoring from change tree. *Proceedings of the First International Workshop on REFactoring, Achievements, Challenges, Effects* (REFACE). Victoria, Canada.

Yu, W., Li, L., et al. (2004). Refactoring use case models on episodes. *Proceedings of the 19th International Conference on Automated Software Engineering* (ASE'04). Linz, Austria, (pp. 328-331).

Yu, Y., Mylopoulos, J., et al. (2003). Software refactoring guided by multiple soft-goals. *Proceedings of the First International Workshop on REFactoring, Achievements, Challenges, Effects* (REFACE). Victoria, Canada.

Zhang, J., & Lin, Y. (2005). Generic and domain-specific model refactoring using a model transformation engine. In *Model-driven software development-research and practice in software engineering*. Springer. doi:10.1007/3-540-28554-7_9

Zimmer, J. A. (2003). Graph theoretical indicators and refactoring. In F. M. a. D. Wells (Ed.), *Proceedings of the Third XP and Segond Agile Universe Conference*. New Orleans, USA, Springer. (pp. 27-53).

ENDNOTES

1. http://swoogle.umbc.edu/
2. http://www.refactoring.com/catalog/index.html
3. http://www.agiledata.org/essays/databaseRefactoringCatalog.html
4. A good class organization in this context means an organization that promotes code reusability, contains a minimum number of multiple and redundant inheritances, and other criteria, depending on the point of view of the author.

[5] http://www.refactorit.com

[6] http://www.togethersoft.com/products/controlcenter/index.jsp

[7] http://www.win.ua.ac.be/~lore/refactoringProject/index.php

[8] http://www.agiledata.org

[9] http://www.refactoring.com/catalog/index.html

[10] http://en.wikipedia.org/wiki/Refactor

Chapter 11
Evolution in Ontology-Based User Modeling

Jairo Francisco de Souza
Federal University of Juiz de Fora, Brazil

Sean Wolfgand Matsui Siqueira
Federal University of the State of Rio de Janeiro, Brazil

Rubens Nascimento Melo
Pontifical Catholic University of Rio de Janeiro, Brazil

ABSTRACT

Web-ontologies are becoming the de facto standard for WWW-based knowledge representation. As a consequence, user modeling has been associated to Web-ontologies. However, data schemes evolve, and therefore ontologies also evolve. Thus, adaptive systems, more than other ontology-based system, are directly affected by changes in ontologies. Because of this, it is important that adaptive systems can be prepared to deal with the problems that occur after changes are applied to ontologies. In this chapter, the authors perform a literature review on the field of ontology evolution aiming at serving as a point of reference for user modeling area. Therefore, adaptive systems developed on ontology-based user modeling could adapt to changes when the ontologies change.

INTRODUCTION

A lot of research activity has been generated on the border of user modeling and web-ontologies. According to Sosnovsky & Dicheva (2010), both disciplines attempt to model the real world phenomena qualitatively: ontologies – a particular area of knowledge, user modeling – the internal state of a human user. Many user-modeling approaches exploit the content-based characteristics of a user (user's knowledge, interests, etc.) and hence can directly benefit of the high-quality domain models provided by ontologies. Besides, as the majority of user modeling projects has been deployed on the Web, and Web-ontologies are becoming the *de facto* standard for WWW-based knowledge representation, the cooperation between user modeling and Web-ontologies seems inevitable (Sosnovsky & Dicheva, 2010).

DOI: 10.4018/978-1-61520-921-7.ch011

During some years, the research about ontologies focused on providing tools for better edit, construct, and visualize these knowledge structures. Nowadays, researchers try to address incompatibility and inconsistence issues that occur after an ontology change, mainly considering that ontologies are part of an environment composed of systems and other ontologies. Thus, this environment must remain functional after some ontology had its structure changed by adding some new knowledge. Adaptive systems, more than any other ontology-based system, are directly affected by changes in the ontologies, as ontologies are increasingly used for model user knowledge and characteristics.

In this chapter, we perform a literature review on the field of ontology evolution that will, hopefully, serve as a point of reference for user modeling area. Our purpose is to give an overview of relevant ontology evolution approaches that can be useful for adaptive system researches and developers. We first make an overview about user modeling and ontology-based user modeling. After that, we discuss the ontology evolution process, its phases and major problems. Then, ontology evolution approaches applied to user modeling area are presented and, finally, we present some final remarks.

USER MODELING

User modeling is mainly used in adaptive systems and in general the precision of modeling assumptions about a user defines, in many aspects, the effectiveness of these systems. An incorrect interpretation of a user leads to wrong adaptive decisions, which may result in user's frustration, loss of trust, decreased motivation to use the system, etc (Sosnovsky & Dicheva, 2010). Crucial factors for the success of adaptive web systems are: adequate representation of knowledge about a user, effective elicitation of user-related information, and utilization of this information for organizing coherent and meaningful adaptation. This section aims to present a brief overview about the user modeling area. For more detailed information, see (Brusilovsky, 1996; Devedzic, 2001; Pierrakos et al, 2003; Stewart et al, 2004; Frias-Martinez et al, 2006; Sosnovsky & Dicheva, 2010).

Several users' characteristics can influence the individual utility of a provided service or information. Some systems model users considering multiple dimensions. Sosnovsky & Dicheva (2010) identify six main dimensions used by user modeling systems over the years (I) knowledge, beliefs, skills, background; (II) interests and preferences; (III) goals, plans, tasks, needs; (IV) demographic information; (V) emotional state and (VI) context. The first dimension is important to information and knowledge systems, which are used for assessing incorrect knowledge or misconceptions (Mabbott et al., 2004), representing procedural knowledge (Corbett & Anderson, 1994), and detailing the relevant experience gained outside the system (so called background knowledge) (Horvitz et al., 1998). The second and third dimensions are most used for recommendation systems such as adaptive recommenders (Pazzani & Billsus, 2007), adaptive search engines (Micarelli et al., 2007) and adaptive browser agents (Lieberman, 1995). The fourth and fifth dimensions can be important in cognitive setting (Desimone, 1999), adapting the system using demographic and emotional characteristics from users (Rodrigo et al., 2007). Demographic characteristics are also used in adaptive e-commerce systems (Bowne, 2000) and personalized ubiquitous applications (Fink & Kobsa, 2002). These last kind of systems can model user contextual information (sixth dimension).

As stated in (Sosnovsky & Dicheva, 2010), there are several approaches to model a user:

- Overlay user modeling is used for modeling user knowledge as a subset of the domain employed by an expert's knowledge. Commonly, this oldest approach is used in adaptive educational systems, as these systems aim to evaluate student's knowledge and to assess how much the student needs to know to become an expert. This approach can also be used for representing user preferences and interests (De Bra et al., 2003).

- Stereotype user modeling (Kay, 1994) is applied when adaptive systems utilize a stock of preset stereotype profiles instead of directly update every single facet of the user model. Whenever a system receives an evidence of a user being characterized by a certain stereotype, the entire user model is updated with the information from this stereotype profile. A user can be described by one stereotype or a combination of several orthogonal stereotypes.
- Keyword-based user modeling aims matching profile models (i.e., user's history of launched queries, rejected emails, accessed documents, etc) against document models (represented as vector of terms – or keywords). This approach is commonly used at information filtering and information retrieval areas (Belkin & Croft, 1992). Many adaptive Web systems model user information interests or needs as vectors of keywords extracted from the documents the user has browsed or requested (Lieberman, 1995; Chen & Sycara, 1998; Billsus & Pazzani, 1999).
- Constraint-based user modeling (Ohlsson, 1996) is used for modeling a user by a set of constraints that represents an acceptable set of equivalent problem statements.
- Collaborative filtering (Burke, 2007) relies on modeling user in terms of their relationships with other users. Typically, a system recommends new items for a user utilizing a vector of ratings the user provided for particular items and matching with a similar vector from a like-minded user.

To populate user information according to the user models, systems generally gather this information from questionnaires or derive it from user's interaction with the system. These systems can use knowledge representation technologies and rely on the semantically-reach overlay or stereotype models, or they can utilize information retrieval techniques and model the user by keywords vectors (Sosnovsky & Dicheva, 2010). Unobtrusive user modeling is preferred on Web systems, due to the exponential growth of information users have to work and users usually do not provide feedback, even if it results in a rewarding behavior of the system. Thus, it is common Web adaptive systems to use machine learning (Webb et al, 2001) and data mining (Frias-Martinez et al, 2006) techniques for content adapting or filtering.

Besides populating user information, it is necessary to consider how to represent the user model. It is important to use mechanisms to allow the computer to process this information. Although relational models are widely disseminated and, thus, provide easier and standard manipulation, other representation models should be considered in order to allow automatic mechanisms for adaptation. In this context, ontologies could be a good solution.

ONTOLOGY-BASED USER MODELING

Ontologies, as a knowledge representation technology, bring benefits for user modeling and adaptation. Even when the major goal for utilizing ontologies is to benefit from one of the derivative ontological technologies (e.g. using ontology mapping for user model mediation or applying ontology-based learning for automatic construction of user model), the primary design decision made by the developer of the adaptive system would be to represent some part of system's knowledge as an ontology (Sosnovsky & Dicheva, 2010). Ontologies can be used by adaptive systems for modeling a domain structure or representing atomic user characteristics based on domain concepts described on the ontology (an approach that inherits overlay user modeling ideas), for structuring a more complex user profile modeling several dimensions of user's state as an ontology (in this case, works using this approach usually encompass stereotype user modeling ideas), or for facilitating keyword-based user modeling using lexical upper ontologies such as WordNet. For instance, Razmerita (2003) proposes an ontology-based user modeling framework and shows how the user ontology can be applied in the context of a Knowledge Management System. Pereira & Tettamanzi (2006) deals with the problem of learning user interests from user behavior in a document retrieval system, in which WordNet (or other lexical database) provides the link between words appearing in a document and the concepts they may refer to (i.e., their meanings). The authors (Pereira & Tettamanzi, 2006) proposed an algorithm that updates the user model as the user interacts with a document retrieval system, discovering the concepts in each document and updating user model interest rates for each concept.

User's characteristics can be modeled using not only ontology concepts but also concept relationships and axioms, improving the information described in the user model (Sicilia et al., 2004). Sosnovsky & Dicheva (2010) exemplifies:

"It could be the knowledge of the fact that a human being 'is-a' mammal (relation) or that every person 'has two and only two' parents (axiom); it can have an interest in anything which is a 'part-of' a particular Laptop model, etc. We do not have knowledge of any system employing such modeling techniques."

In 2007, Dominik Heckmann and colleges revisited the user model and context ontology GUMO (General User Model Ontology) and discussed how Web 2.0 applications can make use of this upper ontology (Heckmann et al, 2007). Several adaptive applications model their users based on GUMO (Kruppa et al., 2005; Stahl et al., 2007; Mori et al, 2007; Neji, 2009). In the same year, Cretton & Calvé (2007) stated they were working in a generic ontology-based user modeling called GenOUM. This ontology would express user's profile, user's knowledge, interests, goals, notifications, evaluations, content's knowledge level, user's behaviors and personality.

Representation of a domain model as an ontology and modeling user's knowledge, interests, needs, or goals as a weighted overlay on the top of it enables the use of standard representation formats and publicly available inference engines, as well as an access to a vast pool of technologies for ontology mapping, quering, learning, etc. User modeling using ontologies are useful for improve searching (Bouzeghoub et al., 2003; Speretta & Gauch, 2005a, 2005b) or to personalize browsing based on user's interests profile (Chaffee & Gauch, 2000; Barla & Bieliková, 2007). Bettina Berendt (2007) discusses some ontological approaches for representing and capturing context.

Standardized domain representation is important for making user models sharable and reusable. However, it's not sufficient. Common vocabularies (such as ontologies) for describing user models are very useful for improving user model reusability and sharability. Design of ontologies describing sharable user profiles (compound knowledge structure reflecting static information about a user, such as user's demography, background, cognitive style, etc.) has become another application of ontologies in user modeling. For instance, Barla & Bieliková (2007) uses an ontological representation of user characteristics because it allows easy interconnection of several models, sharing and reusability of a constructed user model.

Machine learning, data mining, and information storage and retrieval technologies are used for unobtrusive user modeling. Thus, these adaptive systems collecting user information implicitly, in unobtrusive manner, "often face challenges arising due to the conflicts between the computational demands of machine learning and data mining algorithms on one side and the interactive and dynamic nature of user modeling task on the other" (Sosnovsky & Dicheva, 2010; Webb et al., 2001). Interoperability issues and semantic interpretability of user modeling assumptions can make unobtrusive user modeling difficult. Nowadays, research efforts for ontological unobtrusive user modeling demonstrate the potential to partially overcome some of these mentioned difficulties.

The ontology-based unobtrusive user modeling that just adds information in a model (as adding ontology individuals) can be seen as an ontology population task. As Razmerita & Gouardères (2004) stated, ontology-based user modeling requires a referential structure which can be static but also an adaptive part. This structure needs to evolve according to the user's progress, his goals and domains of interest, which need to be acquired and updated. However, as the structure evolves, some inconsistencies can arise in the ontology or interoperability problems issues can occurs (see "Ontology Evolution Process" section). The next section shows how ontology evolution issues are addressed in literature.

ONTOLOGY EVOLUTION PROCESS

Ontologies are not static structures. They hold information regarding a domain of interest, which often changes. Ontologies are used for different purposes and each of them encompasses some need for changing, such as incorporating additional functionalities or knowledge in the ontology, addressing a design flaw in the original conceptualization or a contradiction in the resulting ontology, suiting the ontology with part of other existing ontologies, and so on. Subtle changes in an ontology may have unforeseeable effects in dependent applications, services, data and ontologies (Stojanovic et al. 2002). About this problem, Flouris and his colleges (Flouris et al, 2007) state that *"ontology designers cannot know who uses which part of their ontology and for what purpose, so they cannot predict the effects that a given change on their ontology would have upon dependent elements. The same holds in the opposite direction: if an ontology is depending on other ontologies, there is no way for the ontology designer to control when and how these ontologies will change"*.

Since the problem of ontology evolution is far from trivial (Flouris et al, 2007), it has received a lot of attention in the academic community and it is not near a final solution. However, there is almost a consensus about the phases of the ontology evolution process. Stojanovic et al (2002) studied the ontology evolution process and identified six phases, occurring in a cyclic loop: (1) change capturing; (2) change representing; (3) semantics of change; (4) change implementation; (5) change propagation; and (6) change validation. In the first phase, the changes to be performed are determined. After that, these changes are formally represented. Next, the third phase is the semantics of change phase, in which the effects of the change(s) to the ontology itself are determined; during this phase, possible problems that might be caused to the ontology by these changes are also identified and resolved. So, the changes are physically applied to the ontology, the ontology engineer is informed of the changes and the performed changes are logged. In the propagation phase, these changes are propagated to dependent elements. Finally, the change validation phase allows the ontology engineer to review the changes and possibly undo them, if desired. This last phase may uncover further problems with the ontology, thus initiating new changes that need to be performed to improve the conceptualization (Flouris et al, 2007); in this case, we need to start over by applying the change capturing phase of a new evolution process, closing the cyclic loop. Plessers & de Troyer (2005) identify five phases for ontology evolution, which are very similar to the Stojanovic et al (2002)'s approach.

However, ontology changes can be seen by other aspects. In collaborative environments, for instance, the ontology evolution must lead to specific characteristics. De Leenheer & Mens (2008) works with collaborative ontology change process. They analyze the ontology evolution process for ontologies developed and evolved by a single user and by collaborative and distributed environments. According to the authors, the change process in a collaborative context needs to considerate at least three questions: (1) **What** ontology engineering processes are required in order to achieve the goal or resolve the communication breakdown?; (2) **How** to conduct the activities?; and (3)**Who** will be coordinating these activities? In this context, the ontology evolution process must guarantee communication flow and encompasses argumentation and negotiation tasks for reaching consensus, and integration tasks. In (Souza et al, 2006), such tasks compose a collaborative evolution process called GNoSIS. This process uses well-known business negotiation techniques and concept similarity algorithms for reaching consensus during collaborative ontology integration. Using GNoSIS with an automatic approach within multi-agent systems is proposed in (Souza et al, 2009a). For more information about collaborative ontology evolution, please see (Noy et al, 2006; De Leenheer & Mens, 2008; De Leenheer, 2008).

Other point-of-view of the ontology evolution process can be found in the literature. Philip O'Brien & Syed Abidi (2006), following natural (biological) evolution, propose that ontological evolution is not the process of managing versions of ontologies and their interrelations but instead it is the preservation of beneficial ontology changes over time through the merging and alignment of interoperable ontologies. For the authors, *"this process is not facilitated by users nor is it observed to occur in some predefined, linear order. It occurs "naturally" in that it is gradual, multi-faceted (i.e., influenced by many factors of the environment), and is time-dependent*". Although their point of view is interesting, the authors neither presented advantages of "natural ontology evolution" to real world problems or how ontology engineers can apply the benefits of this process in their systems.

Ontology Evolution Approaches Applied to User Models

The evolution process deals with changes that occur in an ontology after their initial version and the consequences of these changes, such as treating dependent artifacts and keeping the ontology consistence. We consider changes applied in ontologies those that alter the ontology structure (relations, properties or axioms). Ontologies have been used to enrich user models semantically and improve adaptive systems. Several works focus in populating the ontology-based user model and, from the point of view of the process of ontology evolution, ontology evolves as knowledge base but not as a model. The current section presents approaches that aim to evolve the structure of the ontology-based user model, allowing changes in the ontology' structure to reflect the user evolution in an adaptive system. This evolution changes the structure of the ontology (adding new concepts, relationships or axioms) for reflecting new characteristics (knowledge, interests, goals, etc) of the users.

A key issue in developing successful personalized Web applications is to build a user model that accurately represents users' interests. Users' interests frequently change with time. The ability to adapt fast to the current user's interests is an important feature of a user model. Zhang et al. (2007) deal with the problem of modeling Web users by means of a personal ontology. The authors developed SWULPM (Semantic Web Usage Log Preparation Model) that helps recommender systems to integrate usage data and the domain semantics represented by an ontology. Their construction of a user model is based on a semantic representation of the user activity. The user model is constructed unobtrusively by monitoring the user's browsing habits. According to the authors, the evaluation of the recommendation quality based on SWULPM demonstrated an increase of about 10% of accuracy in comparison with recommendations based solely on the usage history taken from shallow sessions.

The systems Quickstep (Middleton et al., 2001) and Foxtrot (Middleton et al., 2003) also update the user ontology describing user interests. These systems use text mining techniques for recommending research papers that matches some of the interests in the user's profile. Zhou et al. (2005) describe a system generating personal "web usage ontologies" represented in OWL and based on transactional logs. They propose a multi-stage mining process using log analysis, fuzzy logic, formal concept analysis, and graph models. Unfortunately, the resulting "web usage ontology" doesn't represent the declarative semantics of a domain. Nevertheless, Zhou et al show how data mining techniques in user activity logs can extract rich models.

Pan et al (2009) proposed a SAM (Spreading Activation Model)-based evolution model of an ontological user model based on behavioral psychology. SAM is a model for learning in associative networks, neural networks or semantic networks (McNamara & Diwadkar, 1996). Their method aims to learn user's interests and to forget past ones. The method utilizes semantic technology to adapt a user model with the support of a domain ontology. User interests are mapped to domain concepts instead of directly generated by analyzing examples. User model is represented in terms of concepts from an ontology a user is interested in, independent of the specific examples of such interests. Examples of user's interests are then classified into the concepts from the ontology and user's interests in such concepts are registered. In this case, the user model proposed by the authors is a subontology, that is, a part of the domain ontology containing only the concepts of the user's interest. Each concept from the user ontology receives a weight. A spreading activation algorithm is used to maintain the interest weights based on the user's ongoing behaviors.

Using ontology-based user models to explicit user's knowledge about a domain and his/her knowledge evolution is a research that is usually applied in the e-learning area. Pérez-Marín et al. (2007) show, with their Willow system, how to learn the conceptual structure of a user model from unstructured text. Willow is an automatic assignment grading system assessing students' free-text answers to teacher's questions. When a student submit an answer to a problem, Willow extracts the concepts covered in the answer and matches them against the concepts learnt from the reference model. As a result, it creates an individual conceptual model of a student reflecting the correctly reported knowledge and possible misconceptions. The evaluation of Willow shows that there is statistically significant correlation between grades given by the systems and actual grades received by the student in class. The OWL-DM system (Denaux et al., 2005) allows user ontology creation using a domain ontology. OWL-OLM is a user modeling component for elicitation of student knowledge by means of interactive dialog. A student is involved in the dialog about the subject domain. Based on student utterances a dialog agent builds a user ontology representing an individual student's conceptualization. The user ontology can extend the domain ontology by expressing relationships that does not exist in the domain ontology. These new relationships reflect a student's understanding of this part of the domain.

Other approaches for evolving ontologies in order to augment the knowledge represented in an ontology are proposed in literature outside e-learning area. These approaches aim at addressing interoperability or communication issues, but present important results that can be applied for user modeling in adaptive systems. Packer et al (2009) propose a technique that enables a software agent to augment its ontology with domain-related concepts by collaborating with other agents, enabling agents to answer queries in a wider range and with more details. An agent responsible for executing queries, when receiving a query with some unknown concept, communicates with domain expert agents for collecting the concept definition and adding it on its own ontology. The authors propose a technique (Packer et al, 2008) that selects fragments from domain expert agents related to the requested concept. These fragments are then incorporated to the ontology of the query agent. Zablith (2009; 2008) proposes the Evolva, a framework that encompasses the whole life cycle of ontologies implemented within NeOn Toolkit editor (Aleksovski et al, 2006). Their framework uses background information, eliminating user interaction from the evolution process. For background information used for mining new knowledge for the ontology, Zablith uses external resources (databases, folksonomies, texts, etc). However, it is not very clear in this work how changes applied in ontology are validated.

By applying changes in an ontology, it's necessary to make sure the resulting ontology and its dependents, such as sub-ontologies and applications, remain operable. Ontology debugging research focuses on the problem that an ontology has to remain consistent under complex changes during evolution. As for some changes there may have several different consistent states of the ontology, Maedche et al (2002) apply the notion of evolution strategy, allowing the user to customize the evolution process according to his/her needs. In (Maedche, 2002), the author specify 16 fine-grained changes that can be performed in the course of ontology evolution (e.g, Add_Concept, Add_Property, Delete_SubConceptOf, Set_Property_Range, etc) and its consequences. These changes can be composed to express changes on a more coarse level (authors specify 12 composite changes). With this change specification, a set of resolution points are obtained by analyzing what consequences each change can have on the ontology according to its definition and dependencies between ontology entities. Each resolution point can also have ontology refinements. The authors suggest a set of ontology refinements induced by analysis of ontology structure, instances, etc, based on heuristics and data mining algorithms (Maedche & Staab, 2001). Since the user can customize the ontology evolution, he/she has the possibility of organizing the ontology in the desired consistent state and to make some additional changes in the ontology that may yield an ontology better suited for the user's needs.

Other works on ontology changes and its consequences are discussed in (Stuckenschmidt & Klein, 2003, Noy & Klein, 2004; Magiridou et al., 2005; Jaziri et al, 2010). Konstantinidis et al (2007) proposed a 5-step workflow which should be followed by an ontology evolution algorithm to determine the side-effects of a change. The authors show that a number of existing ontology evolution algorithms follow this pattern and provide a general formal framework that captures this 5-step pattern. Their ideas can be applied in unobtrusive adaptive systems that aim to evolve a user model structure. A more direct approach to the problem can be found in (Magiridou et al. 2005), where a declarative language for changing (evolving) the data portion of an RDF ontology is introduced; each possible change in this language has a well-defined set of implied side-effects, which are automatically applied by the system. Some authors (Roger et al, 2002; Liu et al, 2006) address the ontology evolution problem with semantics for updating Description Logics theories, since DL is one of the most popular formalisms for ontology representation (Flouris et al, 2007).

During the process of discovering new knowledge or augmenting ontology with new concepts or relations, it's very common to use some algorithms for assessing the similarity of concepts from distinct ontologies. Although there are several approaches for matching concepts in the literature, this problem is far from being fully addressed. Well-known approaches for matching ontologies are implemented in PROMPT (Noy & Musen, 2000) and Chimaera (McGuiness et al,

2000) tools. Cibrasi & Vitányi (2007) propose a similarity measure that utilizes Google search's resulting page counts for assessing concepts' distance based on information distance and Kolmogorov complexity (Li & Vitanyi, 1997). Roughly speaking, the authors use the idea that two concepts are similar as long as they occur simultaneously in the same web pages. A massive randomized trial in binary classification using support vector machines to learn categories based on their Google distance is conducted, resulting in a mean agreement of 87% with the expert crafted WordNet categories. In some cases, it is not sufficient to match concepts from ontologies every time the ontology evolves and warn the ontology dependents about the new ontology version. In not stables scenarios, when ontologies evolve several times, it may be interesting to assess whether the match is stable (robust) or not. Thor et al (2009) use historical information of mappings for assessing if ontology mappings are stable. Their idea is to compare different versions of ontologies and their calculated mappings. When a concept mapping vary a lot between versions, it is possible that this mapping will not exist in future versions, that is, the mapping is unstable. As their techniques use only historical information, it can be applied independently of the match technique that is used. The authors present an evaluation of the approach using two Gene Ontology sub-ontologies. Souza et al (2009b) present a system that uses composition of basic similarity algorithms and apply different weights for each algorithm, showing how different algorithms can be applied together and the importance of an approach to calibrate the weights. These and other similarity algorithms and ontology matching approaches are deeply discussed in (Noy, 2004; Euzenat & Shvaiko, 2004; Euzenat & Shvaiko, 2007).

CONCLUSION

Over the last years, a number of user-adaptive systems have been exploiting ontologies for the purposes of semantics representation, automatic knowledge acquisition, domain and user model visualization and creation of interoperable and reusable architectural solutions (Sosnovsky & Dicheva, 2010). Ontologies need to evolve according to the user's progress in the system, his goals and domains of interest, which need to be acquired and updated. However, as the structure evolves, some inconsistencies can arise in the ontology or interoperability problems issues can occur.

Modern research about ontologies are trying to understand the ontology evolution process and to address the several open issues concerning it, such as, how to discover relevant new knowledge for the ontology, how to describe and apply the changes in the ontology, how to prevent inconsistences on the ontology and its dependents, and how to manage ontology versions.

Issues about ontology evolution influence adaptive systems development, due to (1) these systems can use ontologies as technology for representing user models and, because that, they must evolve the ontology always the user model needs to evolve; or (2), since these systems use an ontology, either to understand the domain knowledge or to address interoperability issues, the adaptive system is dependent of some ontology, that is, it is part of the ontology evolution process.

In this chapter, the ontology evolution was presented and discussed. Moreover, several approaches concerning the evolution of ontologies were presented in order to contribute to adaptive systems development and research. For more information about ontology evolution issues and state-of-art, see (Flouris et al, 2007; De Leenheer & Mens, 2008).

REFERENCES

Aleksovski, Z., Klein, M., Kate, W., & Van Harmelen, F. (2006). Matching unstructured vocabularies using a background ontology. In Staab, S., & Svátek, V. (Eds.), *EKAW 2006. (LNCS/LNAI 4248)* (pp. 182–197).

Barla, M., & Bieliková, M. (2007). Estimation of user characteristics using rule-based analysis of user logs. *Proceedings of Data Mining for User Modeling Workshop held at the International Conference on User Modeling,* (pp. 5-14).

Belkin, N. J., & Croft, W. B. (1992). Information filtering and information retrieval: Two sides of the same coin? *Communications of the ACM, 35*(12), 29–38. doi:10.1145/138859.138861

Berendt, B. (2007). Context, (e)learning, and knowledge discovery for Web user modelling: Common research themes and challenges. *Proceedings of Data Mining for User Modeling Workshop held at the International Conference on User Modeling,* (pp. 15-27).

Billsus, D., & Pazzani, M. J. (1999). A hybrid user model for news story classification. In *Proceedings of the 7th International Conference on User Modeling* (UM'1999), (pp. 99- 108), Banff, Canada: Springer-Verlag, New York, Inc.

Bouzeghoub, A., Carpentier, C., Defude, B., & Duitama, J. F. (2003). A model of reusable educational components for the generation of adaptive courses. In *Proceedings of the First International Workshop on Semantic Web for Web-Based Learning at CAISE'03,* Klagenfurt, Austria. Retrieved from http://www-inf.int-evry.fr/~defude/PUBLI/SWWL03.pdf

Bowne. (2000). *Open sesame 2000.* Retrieved from http://www.opensesame.com

Brusilovsky, P. (1996). Methods and techniques of adaptive hypermedia. *User Modeling and User-Adapted Interaction, 6*(2-3), 87–129. doi:10.1007/BF00143964

Burke, R. (2007). Hybrid Web recommender systems. In Brusilovsky, P., Kobsa, A., & Nejdl, W. (Eds.), *The adaptive Web: Methods and srategies of Web personalization* (pp. 377–408). Heidelberg, Germany: Springer Verlag.

Chaffee, J., & Gauch, S. (2000). Personal ontologies for Web navigation. In *Proceedings of the 9th International Conference on Information and Knowledge Management* (CIKM '2000), (pp. 227-234). McLean, VA/ New York, NY: ACM.

Chen, L., & Sycara, K. (1998). A personal agent for browsing and searching. In *Proceedings of the 2nd International Conference on Autonomous Agents* (pp. 132-139). Minneapolis/St. Paul.

Cilibrasi, R., & Vitányi, P. (2007). The Google similarity distance. *IEEE Transactions on Knowledge and Data Engineering, 19*(3), 370–383. doi:10.1109/TKDE.2007.48

Corbett, A. T., & Anderson, J. R. (1994). Knowledge tracing: Modeling the acquisition of procedural knowledge. *User Modeling and User-Adapted Interaction, 4*(4), 253–278. doi:10.1007/BF01099821

Cretton, F., & Calvé, A. L. (2007). *Working paper: Generic ontology based user modeling-GenOUM.* (Technical report, HES-SO Valais).

De Bra, P., Aerts, A., Berden, B., de Lange, B., Rousseau, B., Santic, T., et al. (2003). AHA! The adaptive hypermedia architecture. In *Proceedings of the Fourteenth ACM Conference on Hypertext and Hypermedia* (pp. 81-84). Nottingham, UK: ACM Press.

De Leenheer, P. (2008). Keynote: Towards an ontological foundation for evolving agent communities. *Proceedings of 32nd Annual IEEE International Computer Software and Applications Conference,* (pp. 523-528).

De Leenheer, P., & Mens, T. (2008). Ontology evolution: State of the art & future directions. In Hepp, M., De Leenheer, P., de Moor, A., & Sure, Y. (Eds.), *Ontology management for the Semantic Web, Semantic Web services, and business applications, from Semantic Web and beyond: Computing for human experience*. Springer.

Denaux, R., Aroyo, L., & Dimitrova, V. (2005a). An approach for ontology-based elicitation of user models to enable personalization on the Semantic Web. In *Proceedings of the 14th international conference on World Wide Web* (WWW '05), (pp. 1170-1171), Chiba, Japan. New York, NY: ACM.

Desimone, L. (1999). Linking parent involvement with student achievement: Do race and income matter? *The Journal of Educational Research, 93*(1), 11–30. doi:10.1080/00220679909597625

Devedzic, V. (2001). Knowledge modeling-state of the art. *Integrated Computer-Aided Engineering, 8*(3), 257–281.

Euzenat, J., & Shvaiko, P. (2004). A survey of schema-based matching approaches. *Journal on Data Semantics, 4*, 146–171.

Euzenat, J., & Shvaiko, P. (2007). *Ontology matching*. Berlin/ Heidelberg, Germany: Springer-Verlag.

Fink, J., & Kobsa, A. (2000). A review and analysis of commercial user modeling servers for personalization on the World Wide Web. *International Journal on User Modeling and User-Adapted Interaction, 10*(2-3), 209–249. doi:10.1023/A:1026597308943

Fink, J., & Kobsa, A. (2002). User modeling in personalized city tours. *Artificial Intelligence Review, 18*(1), 33–74. doi:10.1023/A:1016383418977

Flouris, G. (2007). Ontology change: Classification and survey. [United Kingdom: Cambridge University Press.]. *The Knowledge Engineering Review, 23*(2), 117–152.

Frias-Martinez, E., Chen, S., & Liu, X. (2006). Survey of data mining approaches to user modeling for adaptive hypermedia. *IEEE Transactions on Systems, Man and Cybernetics. Part C, Applications and Reviews, 36*(6), 734–749. doi:10.1109/TSMCC.2006.879391

Heckmann, D., Schwarzkopf, E., Mori, J., Dengler, D., & Kröner, A. (2007). The user model and context ontology GUMO revisited for future Web 2.0 extensions. *Proceedings of Contexts and Ontologies: Representation and Reasoning*, (pp. 37-46).

Horvitz, E., Breese, J., Heckerman, D., Hovel, D., & Rommelse, D. (1998). The Lurniere project: Bayesian user modeling for inferring the goals and needs of software users. In *Proceedings of the 14th Conference on Uncertainty in Artificial Intelligence*, (pp. 256-265), Madison, WI: Morgan Kaufmann Publishers.

Kay, J. (1994). Lies, damned lies and stereotypes: Pragmatic approximations of users. In A. Kobsa & D. Litman (Eds.), *Proceedings of the Fourth International Conference on User Modeling* (UM'1994), (pp. 175-184). Hyannis, MA: MITRE Corp.

Konstantinidis, G., Flouris, G., Antoniou, G., & Christophides, V. (2007). Ontology evolution: A framework and its application to RDF. *Proceedings of the Joint ODBIS & SWDB Workshop on Semantic Web, Ontologies, Databases* (SWDB-ODBIS-07).

Kruppa, M., Heckmann, D., & Krueger, A. (2005). Adaptive multimodal presentation of multimedia content in museum scenarios. *KI - Zeitschrift fuer Kuenstliche Intelligenz, 5*(1), 56-59.

Li, M., & Vitanyi, P. (1997). *An introduction to Kolmogorov complexity and its applications*. New York, NY: Springer-Verlag.

Lieberman, H. (1995). Letizia: An agent that assists web browsing. In *Proceedings of the International Joint Conference on Artificial Intelligence*, (pp. 924–929). Montreal, Canada.

Liu, H., Lutz, C., Milicic, M., & Wolter, F. (2006). Updating description logic ABoxes. *Proceedings of the 10th International Conference on Principles of Knowledge Representation and Reasoning* (KR-06).

Mabbott, A., & Bull, S. (2004). Alternative views on knowledge: Presentation of open learner models. In J. C. Lester, R. M. Vicari & F. Paraguacu (Eds.), *Proceedings of the 7th International Conference on Intelligent Tutoring Systems* (ITS'2004), (pp. 689-698). Berlin/ Heidelberg, Germany: Springer-Verlag.

Maedche, A. (2002). *Ontology learning for Semantic Web*. Kluwer Academic Publishers.

Maedche, A., Motik, B., Stojanovic, L., & Stojanovic, N. (2002) User-driven ontology evolution management. *Knowledge Engineering and Knowledge Management: Ontologies and the Semantic Web, 2473*(1), 133-140. Heidelberg, Germany: Springer-Verlag.

Maedche, A., & Staab, S. (2001). Ontology learning for the Semantic Web. *IEEE Intelligent Systems, Special Issue on Semantic Web, 16*(2).

Magiridou, M., Sahtouris, S., Christophides, V., & Koubarakis, M. (2005). RUL: A declarative update language for RDF. *Proceedings of the 4th International Semantic Web Conference* (ISWC-05), (pp. 506-521).

McGuiness, D., Fikes, R., Rice, J., & Wilder, S. (2000). An environment for merging and testing large ontologies. *Proceedings of the 7th International Conference on Principles of Knowledge Representation and Reasoning* (KR-00), (Technical Report KSL-00-16, Knowledge Systems Laboratory, Stanford University).

McNamara, T. P., & Diwadkar, V. A. (1996). The context of memory retrieval. *Journal of Memory and Language*, *35*(1), 877–892. doi:10.1006/jmla.1996.0045

Micarelli, A., Gasparetti, F., Sciarrone, F., & Gauch, S. (2007). Personalized search on the World Wide Web. In Brusilovsky, P., Kobsa, A., & Nejdl, W. (Eds.), *The adaptive Web: Methods and strategies of Web personalization* (pp. 195–230). Heidelberg, Germany: Springer Verlag.

Middleton, S. E., De Roure, D. C., & Shadbolt, N. R. (2001). Capturing knowledge of user preferences: Ontologies in recommender systems. In *Proceedings of the 1st International Conference on Knowledge Capture* (K-CAP '01), (pp. 100-107), Victoria, British Columbia, Canada. New York, NY: ACM Press.

Middleton, S. E., Shadbolt, N. R., & De Roure, D. C. (2003). Capturing interest through inference and visualization: ontological user profiling in recommender systems. In *Proceedings of the 2nd International Conference on Knowledge Capture* (K-CAP '03), (pp. 62-69), Sanibel Island, FL, USA. New York, NY: ACM Press.

Mori, J., Heckmann, D., Matsuo, Y., & Jameson, A. (2007). Learning ubiquitous user models based on users' location history. *Proceedings of Data Mining for User Modeling Workshop held at the International Conference on User Modeling*, (pp. 40-49).

Neji, M. (2009). The innovative approach to user modelling for information retrieval system. *Proceedings of the 20th ACM Conference on Hypertext and Hypermedia.*

Noy, N. (2004). Semantic integration: A survey of ontology-based approaches. *SIGMOD Record*, *33*(4), 65–70. doi:10.1145/1041410.1041421

Noy, N., Chugh, A., Liu, W., & Musen, M. (2006). A framework for ontology evolution in collaborative environments. *Proceedings of 5th International Semantic Web Conference*, (pp. 544–558).

Noy, N., & Klein, M. (2004). Ontology evolution: Not the same as schema evolution. *Knowledge and Information Systems, 6*(4), 428-440. (SMI technical report SMI-2002-0926).

Noy, N., & Musen, M. (2000). Algorithm and tool for automated ontology merging and alignment. *Proceedings of the 17th National Conference on Artificial Intelligence* (AAAI-00), (SMI technical report SMI-2000-0831).

O'Brien, P., & Abidi, S. (2006). Modeling intelligent ontology evolution using biological evolutionary processes. *Proceedings of IEEE International Conference on Engineering of Intelligent Systems,* (pp. 1-6).

Ohlsson, S. (1996). Learning from performance errors. *Psychological Review, 103*(2), 241–262. doi:10.1037/0033-295X.103.2.241

Packer, H., Gibbins, N., & Jennings, N. (2009). Ontology evolution through agent collaboration. *Proceedings of the Workshop on Matching and Meaning 2009: Automated Development, Evolution and Interpretation of Ontologies.*

Packer, H., Gibbins, N., Payne, T., & Jennings, N. (2008) Evolving ontological knowledge bases through agent collaboration. *Proceedings of the Sixth European Workshop on Multi-Agent Systems.*

Pan, J., Zhang, B., & Wu, G. (2009). A SAM-Based evolution model of ontological user model. *Proceedings of the 2009 Eigth IEEE/ACIS International Conference on Computer and Information Science,* (pp. 1139-1143).

Pazzani, M. J., & Billsus, D. (2007). Content-based recommendation systems. In Brusilovsky, P., Kobsa, A., & Nejdl, W. (Eds.), *The adaptive Web: Methods and strategies of Web personalization* (pp. 325–341). Heidelberg, Germany: Springer Verlag.

Pereira, C., & Tettamanzi, A. (2006). An evolutionary approach to ontology-based user model acquisition. *Fuzzy Logic and Applications, 2955*(1), 25-32. Heidelberg, Germany: Springer Verlag.

Pérez-Marín, D., Alfonseca, E., & Pascual-Nieto, I. (2007). Automatic generation of students' conceptual models from answers in plain text. In C. Conati, K. McCoy & G. Paliouras (Eds.), *Proceedings of the 11 International Conference on User Modeling* (UM'2007), (pp. 329-333). Corfu, Greece. Berlin / Heidelberg, Germany: Springer.

Pierrakos, D., Paliouras, G., Papatheodorou, C., & Spyropoulos, C. D. (2003). Web usage mining as a tool for personalization: A survey. *User Modeling and User-Adapted Interaction, 13*(4), 311–372. doi:10.1023/A:1026238916441

Plessers, P., & de Troyer, O. (2005). Ontology change detection using a version log. *Proceedings of the 4th International Semantic Web Conference* (ISWC-05), (pp. 578-592).

Razmerita, L., & Gouardères, G. (2004). Ontology based user modeling for personalization of Grid learning services. *Proceedings of the Grid Learning Services Workshop (GLS 2004) in Association with Intelligent Tutoring System Conference.*

Rodrigo, M., Baker, R., Maria, L., Sheryl, L., Alexis, M., Sheila, P., et al. (2007). Affect and usage choices in simulation problem solving environments. In R. Luckin, K. R. Koedinger & J. Greer (Eds.), *Proceedings of the 13th International Conference on Artificial Intelligence in Education* (AIED'2007), (pp. 145-152), Marina Del Ray, CA, USA. IOS Press.

Roger, M., Simonet, A., & Simonet, M. (2002). Toward updates in description logics. *Proceedings of the 2002 International Workshop on Description Logics* (DL-02), CEUR-WS 53(1).

Sicilia, M.-A., Garcia, E., Diaz, P., & Aedo, I. (2004). Using links to describe imprecise relationships in educational contents. *International Journal of Continuing Engineering Education and Lifelong Learning, 14*(3), 260–275. doi:10.1504/IJCEELL.2004.004973

Sosnovsky, S., & Dicheva, D. (2010). Ontological technologies for user modeling. *International Journal of Metadata. Semantics and Ontologies*, *5*(1), 32–71. doi:10.1504/IJMSO.2010.032649

Souza, J. F., Melo, R. N., Oliveira, J., & Souza, J. M. (2009b). Combining resemblance functions for ontology alignment. *Proceedings of the 11th International Conference on Information Integration and Web-based Applications & Services* (iiWAS2009).

Souza, J. F., Paula, M., Oliveira, J., & Souza, J. M. (2006). *Meaning negotiation: Applying negotiation models to reach semantic consensus in multidisciplinary teams. Proceedings of Group Decision and Negotiation* (pp. 297–300). Karlsruhe: Universitätsverlag Karlsruhe.

Souza, J. F., Siqueira, S. W. M., & Melo, R. N. (2009a). Adding meaning negotiation skills in multiagent systems. *Proceedings of IEEE International Conference on Intelligent Computing and Intelligent Systems*, (pp. 663-667). Shangai, China: IEEE Press.

Speretta, M., & Gauch, S. (2005a). Misearch. In *Proceedings of the The IEEE/WIC/ACM International Conference on Web Intelligence 2005,* (pp. 807- 808). Chicago, IL, USA.

Speretta, M., & Gauch, S. (2005b). Personalized search based on user search histories. In *Proceedings of the The IEEE/WIC/ACM International Conference on Web Intelligence 2005,* (pp. 391-398). Chicago, IL, USA.

Stahl, C., Heckmann, D., Schwartz, T., & Fickert, O. (2007). Here and now: A user-adaptive and location-aware task planner. In S. Berkovsky, K. Cheverst, P. Dolog, D. Heckmann, T. Kuflik, P. Mylonas, … J. Vassileva (Eds.), *Proceedings of the Workshop on Ubiquitous and Decentralized User Modeling (UbiDeUM) at UM'200,7* (pp. 52-67).

Stewart, A., Niederée, C., & Mehta, B. (2004). *State of the art in user modelling for personalization in content, service and interaction.* NSF/DELOS Report on Personalization.

Stojanovic, L., Maedche, A., Motik, B., & Stojanovic, N. (2002). User-driven ontology evolution management. *Proceedings of the 13th International Conference on Knowledge Engineering and Knowledge Management* (EKAW-02), (LNCS 2473), (pp. 285-300). Heidelberg, Germany: Springer-Verlag.

Stuckenschmidt, H., & Klein, M. (2003). Integrity and change in modular ontologies. *Proceedings of the 18th International Joint Conference on Artificial Intelligence* (IJCAI-03). Jaziri W., Sassi N., & Gargouri F. (2010). Approach and tool to evolve ontology and maintain its coherence. *International Journal of Metadata, Semantics and Ontologies.* United Kingdom: Inderscience Publishers.

Thor, A., et al. (2009). An evolution-based approach for assessing ontology mappings-a case study in the life sciences. *Proceedings of the 13th Conf. of Database Systems for Business, Technology and Web* (BTW).

Webb, G. I., Pazzani, M. J., & Billsus, D. (2001). Machine learning for user modeling. *International Journal of User Modeling and User-Adapted Interaction*, *11*(1-2), 19–29. doi:10.1023/A:1011117102175

Zablith, F. (2008). Dynamic ontology evolution. *Proceedings of the International Semantic Web Conference (ISWC)* Doctoral Consortium. Karlsruhe, Germany.

Zablith, F. (2009). Ontology evolution: A practical approach. *Proceedings of the AISB Workshop on Meaning and Matching* (WMM), Edinburgh, UK.

Zhang, H., Song, Y., & Song, H.-t. (2007). Construction of ontology-based user model for Web Personalization. In C. Conati, K. McCoy & G. Paliouras (Eds.), *Proceedings of the 11 International Conference on User Modeling* (UM'2007), (pp. 67-76), Corfu, Greece. Berlin /Heidelberg, Germany: Springer.

Chapter 12
Collaborative Filtering:
Inference from Interactive Web

Tania Al. Kerkiri
University of Macedonia, Greece

Dimitris Konetas
University of Ioannina, Greece

ABSTRACT

The interactive tools like blogs, wikis, et cetera, known under the commonly acceptable name Web2.0, led to a new generation of Internet services and applications such as social networks, recommendation systems, reputation systems, et cetera, allowing for public participation in the formation of the content of the Web, and at the same time fueling an explosion of information. This information is a widely available intellectual capital. Due to the opportunities that arise from the exploitation of this information, this chapter will i) present the rationale under which these systems function, ii) summarize and apply in an indicative manner the mathematical models used to handle this information, iii) propose a general architecture of these systems and iv) describe a hybrid multifaceted algorithm that exploits the capabilities arising from this information towards a personalized inference for a specific user. The result of this work is an indication of the capabilities that arise from further exploitation of these systems.

1. INTRODUCTION

Interactive web applications appear to be the relatively novel modus vivendi for a number of everyday activities and at the same time the novel set of increasing demands for improvements. The available tools for these applications create virtual communities between their users. General categories of these applications include, inter alia: Instant Messaging, Text chat, Internet forums, Wikis, Blogs, Collaborative real-time editors, Social network services, Commercial social networks, Social bookmarking, Social cataloging, non-game worlds (with most indicative among them: Second Life, Crocquet, The Sims Online), specialized social applications, peer-to-peer social

DOI: 10.4018/978-1-61520-921-7.ch012

networks, etc. An analytical description as well as several examples of these may be found in <http:// en.wikipedia.org/wiki/ Social_software>. These applications, widely known under the term Web 2.0 and/or Enterprise 2.0, share characteristics, service-oriented design and the ability to upload data and media. The successful handling of this information is vital – if information treatment is to reach its new challenging era of a novel quality regarding communicative affairs. Upendra (1995) shown that a prominent tactic for its exploitation is collaborative filtering methods (a collection of their features can be found in <http://en.wikipedia. org/wiki/Collaborative_ filtering>), which may sufficiently be used to make this information widely useful. As shown in Schafer (2007) the huge amount of information created by these applications find specialized usage in e-commerce systems. Resnick (1997) shown another usage of them in recommendation systems and reputation systems (details of their function can be found in <http://en.wikipedia.org/wiki/Reputation_system>), censorship systems, digital libraries and lately in social networks, e.g. in Parra (2009) and Ziegler (2004), where a research has been done on systems that handle this kind of information for the social tools, and systems that fulfill the semantic web technologies. However, new constructs might as well ameliorate these approaches: In this article **i)** the philosophy of such systems is presented along with various mathematical methods that exploit the relevant available information, through an example borrowed from the e-Learning area, **ii)** their functional architecture is sketched and **iii)** the novel construct of a hybrid algorithm is described that highlights several options of how this information can be used with a view to pertinent predictions for serving the individual needs of specific users. Finally, conclusions are drawn with respect to the variants of such a construct combined with the indicative methodology presented toward optimally leading to a general model for managing the information of similar systems, in anticipation for the improvement of their functionality.

2. COLLABORATIVE-FILTERING METHODS: WHAT ARE THEY?

Balabanović (1997) declares collaborative-based filtering methods as algorithms that trace items for the user who interacts with the system at any given time, based on information collected from users which have experienced the system up to now (Resnick, 2000). The idea underlying the operation of these systems is based on two basic assumptions:

- that people who have agreed in the past on a given issue tend to agree again in the future. For example, if people rated as 'Interesting', 'Useful', or 'Good' the same items at the same level, then they are likely to express the same opinion for similar items which might be presented to them in the future, and
- that people having similar characteristics tend to make the same choices.

The usual usage of the systems applying collaborative filtering algorithms includes:

- the promotion of products / points of view,
- the creation of bond of trust between people that participate in digital communities and meet each other only electronically,
- the support of the reputation of e-commerce practitioners,
- the decision about the strategy through which the items of a collection are selected and presented to the user by affecting their order.

A number of web-applications apply more or less CF-methods nowadays. The results of this application can be seen as: '...users that have seen this item they also bought the following...', or '... tell us your opinion about...', or even in 'Similar pages' which is found near web-pages retrieved from a search engine like Google.

Generally, CF-algorithms work in a designated manner: they collect information from the navigation of the users in the web-pages, they mediate on this information and exploit its accumulated results so as to group these users into similarity sets ('neighborhoods'). After that, inference for a specific user is made using the information they have collected from his/her neighbors. As Herlocker (2000) noted, the CF-algorithms do not search for items, as the usual search systems do by retrieving items accordingly to their actual properties/content. Instead, they *predict* items for the current user based on feedback provided from users who experienced the same items in the past. Actually, the systems that embed CF-algorithms must have collected a great amount of historic data from, and for their users to be able to provide reliable results.

Let's call the systems that embed CF-algorithms, CF-systems. These systems collect information for their users through two main ways:

- tracking information from their navigation –this information being called 'indirect' evaluation, and
- collecting ranking for items/products/ideas/services, asking for, and inducing the users' intentional evaluation –these being 'direct' evaluation.

Both types of information concern a number of things, e.g. visited pages, items have been bought, time spent over a page, people linked to a profile, etc. The indirect information is not totally reliable, but it has the advantage that its collection does not impose the involvement of the user.

On the other hand, the direct ranking is of great importance for these systems, as: **i)** it comes from a person that had a real experience with the item, and **ii)** the person was interested for the item s/he ranked. For these reasons several fairly aggressive practices are used to collect direct ranking, such as:

- the transaction is not allowed to be completed unless users provide their ranking,
- the users are induced to rank, promising that, through their contribution, the services of the system will be improved,
- the users are rewarded for their rank, e.g. by being offered discount coupons for next transactions.

As the main outcome of CF-systems is the matching of two entities, e.g. products and consumers, employers-jobs, friends, etc, these systems must maintain a representation of the characteristics of the items they promote and a representation of the characteristics of their users. So, let $I = \{i_1, i_2, ..., i_m\}$ be the set of the items they handle. For each instance i_l of these items a set of characteristics/properties is defined, and let $i_l = \{p_1, p_2, ..., p_l\}$ be the set of these properties for item i_l. These properties vary depending on the system that are applied; e.g., **i)** for a product, they may be: color, price, quality, response-time in its dispatching, etc, and **ii)** for an educational resource, they can be: comprehensibility, originality, conciseness, objectives, background of the learner to whom they are addressed, etc. (Kerkiri, in press).

Similarly, these systems define the set $C = \{c_1, c_2, ..., c_n\}$ of their users, and let c_j, j=1,..,n, be an individual user. For each instance c_j the vector $c_j = \{p_1, p_2, ..., p_k\}$ depicting the user's properties is defined. Even these properties vary depending on the case: **i)** for a commercial system these properties may be: immediate / safe dispatching of the product, the fidelity of both parties to the terms of the agreement, etc, while **ii)** in an electronic educational system may be: the user's previous expertise / background, grades, portfolio of lessons / examinations / participations in educational activities, age, sex, etc.

These vectors of the properties are ordered, so each of their elements always represents the same property.

The information about the users constructs their profile. This profile is built gradually by attaching to the vector a great number of information collected from their everyday interaction with the system. Notably this information may also be gathered **i)** ***directly***, e.g. the information that a user provides during his registration to a website, or **ii)** ***indirectly***. This latter kind of collection exploits the current capabilities of network and web applications: e.g. through these capabilities the IP and GIS axis of the users are available; consequently, the zone they enter the system and a number of --geographic details. Every piece of information is precious, e.g., demographic evidence on marital status, place of living, preferences, financial information, the visited items, the overall remaining time over an item, the last time they visited the system, the number of returning times to a specific site, the frequency of returning and so on.

In the following section the operation of these systems is described through an example borrowed from the e-Learning field. As mentioned in Kerkiri (2010) e-Learning is an area that can very profitably apply CF-algorithms, as significant information for their (users)learners is available.

There are two main reasons for that:

- typical Learning Management Systems (LMSs) may maintain thousands of locally-stored courses and links to an exponentially increasing amount of data. Itmazi (2008) claims that this overload of information is a potential to improve new knowledge but can equally is a danger of providing no articulate knowledge actually choking with unstructured information the educational system.
- in such systems multiple information is gathered in the learners' profile, using both direct and indirect recording. Direct information is stored by the secretary office of the institution the learner attends and it is also updated from his/her everyday participation in the educational life. Indirect information is gathered by his/her navigation in the electronic media which is used to provide the courses. Consequently, in e-Learning systems a variety of CF-based methods can be combined in order to detect similarity between learners.

Finally, the functional architecture of a system that applies CF-methods is described; align with a hybrid algorithm that outlines a number of strategies that can be applied to find out resources and activities for a specific learner intending to facilitate/improve his learning

3. AN EXEMPLAR OF COLLABORATIVE FILTERING APPLICATION IN E-LEARNING

To depict the functionality of these algorithms in an e-Learning system, let the user of an e-Learning system, Learner for the rest of the paper, and the items he uses, called Learning resources, LRs, for now on. In Kritikou (2008) it is mentioned that in the Learner's profile the information being gathered usually consists of: age, skills, existing certification and background, the courses he attended, the performance/grades he achieved, the preferred level of difficulty of the learning material, the behavior in the e-classroom, the preferred language and format (e.g. doc, pdf, avi) the books he studies must have, the educational activities he participates etc.

To exploit this information

- a weight must be assigned to each of these properties,
- the permitted values of each property must be defined along with
- a weight for each of these values. The representation of the properties with an arith-

metic value is necessary in order to apply mathematical models for computing the similarity between the learners.

An example borrowed from an e-Learning environment follows. In Kerkiri (2009) the application of this system has shown promising outcomes. The properties being examined in this system, along with their permitted values, are:

These weights are indicative; they depend on the desired functionality and the willing outcomes of the system.

The weight also attached to each of these properties is proportional to the desirable effect of the specific property in the system being applied. As an example, let's consider the previous education of the learner and mark it as of great importance, so the property 'Certification' is assigned a weight of 0.9; similarly thinking the property 'Role' is assigned a weight of 0.5.

Having this information, the vector depicting the properties of the profile of a learner is:

```
ci={Age, Certifications, Role, Level, Lessons, PreferredLanguage, PreffeferedForrmat}
```

We can see the information attached to a learner's profile: the websites he visited, the books he read etc, as well as the action he performed in each of them. As we can see from Example 1 the actions can either be 'online study', 'download' for further study, or 'ranking'. The LRs marked as 'ranked', from those that are attached to a learner's profile, have 'direct' evaluation, the others are supposed to have 'indirect' ranking.

The combination of the weights of the properties and the weights of the values of each property gives a total value for each Learner. E.g. for a specific learner who has a 'MSc' degree, 'Moderate participation', is 'Intermediate+', and has average 'Performance' 0.7 his total weight is:

```
ci ={ 0.9*Certification, 0.5*Role, 0.6*Level, 0.7*Performance}=> ci ={
```

Table 1.Properties, permitted values and indicative weights for attributes for a learner

Property	Property Weight	Values	Property Values Weight
Certifications	0.9	High School Level/Vocational Education	0.2
		Technological Institute Education	0.4
		Higher Education	0.6
		MSc	0.8
		PhD	1
Level	0.6	Beginner	0.2
		Intermediate-	0.4
		Intermediate	0.6
		Intermediate+	0.8
		Advanced	1
Performance	0.7	Analytical portfolios from the courses and the events he attended and, finally, the mean grades s/he accumulated	
Role in educational life	0.5	Indifferent	0.2
		Moderate participation	0.4
		Sufficient	0.6
		Important	0.8
		Dynamic	1

Box 1.

```
ci ={Age, Certifications, Role, Level, {Title, Avg,{Activities}, {LRs} }, Preferred-
Language, PrefferedForrmat}=
ci ={Age, Certifications, Role, Level, {Title, Avg,
                                         {Title of the Activity, Grade},
                                         {Title of the LR, Format, Action} },
                                                PreferredLanguage, PrefferedFor-
rmat}
```

```
0.9*MSc, 0.5* Moderate participation,
0.6* Intermediate+, 0.7*Performance
}=> ci ={ 0.9*0.8, 0.5*0,6, 0.6* 0.8,
0.7*0.7 }=> ci ={ 0.72+0.30+0.48+0.49
}=1.99
```

This number has a dual role: **i)** it can be exploited by mathematical models, **ii)** it multiplies each rank a learner provides for the items he visits, in other words, it is the weight through which the learner affects the system.

A similar vector is created for each participant of the system.

A method is needed to find out the similarity between two learners align with the values of the properties of their profile. These methods are based

Box 2. Example 1: The profile of an IT-learner

```
ci={25, Higher Education, Moderate participation, Intermediate+,
    { {Databases II, Avg=7.56, Exersice1={title='Development of a database for a
health system', grade=8.5},
                              Exersice2={title='System of remote power management',
grade=6.4},
                              Final Examinations, grade=7.8},
                              {1. 'Databases II', Book Ed. XYZ, action: download,
                               2. 'Databases and web-applications', Html page, ac-
tion: view,
                               3. Booklet, set of exercises, action: rank=0.315} },
    {Informational systems, Avg=8.5, Exerise1={title='', grade=null},
                                    Examination=8.5},
                              {1. Instructor's manual, book, action: rank=0.526} },
    {Algorithm, Avg=8.3,   Exersice1={'Algorithms for traffic regulation of Piraeus
Str. lights', 8.1},
                                    Examination=8.6},
                              {1. Algorithms, Book Ed. Knuth, action: download,
                               2. Instructor's book, Book, action: tank=0.615,
                               3. 'Algorithm theory', website, action: visit}},
    },
  {en, el}, {doc, pdf}
}
```

on the hypothesis that similar users tend to make similar choices. To implements such methods in Kerkiri (in press), the following definition is provided:

Similarity Definition 1. *Two Learners ci and cj are similar when the distance between the values of their properties is less than a pre-defined threshold.*

First of all, the meaning of 'how close two learners are' must be defined. In a perfect situation the two vectors will fit exactly. Actually, this is not feasible, not even desirable. Generally, in these systems it is desirable to increase the similarity sets, by adding Learners that could probably match in the future. So, we have to find the maximum distance over which two Learners are no longer similar. In Salton (1975) this distance is defined as the 'threshold' of the system and several methods exist for its computation.

The Vector Space Model is a good proposal to compute the learners' similarity from the matching of their properties. The VSM computes the distance of two vectors based on the inner product of their elements (formula 1, shown in Box 3), where c_i, c_j are the vectors of the two Learners, and w_{ik}, w_{jk} are their weights–as they were calculated from the arithmetic values of the properties of their profile.

If the two vectors do not have the same number of values, then 0s are added in the place of the missing ones. Actually, the VSM represents the cosine of the angle between the two vectors. Formula (1) derives values from 0-1. Consequently, the closer to 1 the outcome of this formula is, the more similar the items are.

To find out if two learners belong in the same neighborhood the threshold by which their similarity is compared must be defined. The method used here to find out the threshold depends on the data set being examined. So, the threshold is defined as the distance from the unit, of the average similarities that derive of the learners compared in pairs (formula 2).

$$threshold_p = 1 - \frac{\sum_{j=i+1}^{n} sim(c_i, c_j)}{n * (n-1)} \quad (2)$$

Consequently, if the result of each comparison is less than the threshold, the two learners are members of the same neighborhood.

A second way to find out similarity between learners is based on the learners' evaluations for the items they ranked. Methods of this kind are based on the hypothesis that users that evaluated similarly the same items in the past perhaps they will meet again in the future. To provide a method that uses these evaluations the following definition is defined:

Similarity Definition 2. *Similar learners are those that the value of the correlation of the evaluations they provided for a set of commonly used LRs is less that a threshold.*

The mathematical model used here to combine the direct ranking is the Pearson's correlation model. In several research, such as (Tsai, 2006;

Box 3.

$$sim(c_j, c_i) = \frac{\vec{c_j} * \vec{c_i}}{|\vec{c_j}| * |\vec{c_i}|} = \frac{\sum_{k=1}^{t}(w_{j,k} * w_{i,k})}{\sqrt{\sum_{k=1}^{t} w_{j,k}^2} * \sqrt{\sum_{k=1}^{t} w_{i,k}^2}} \leq threshold_p \quad (1)$$

Wang, 2007; Kerkiri, 2009), a lot of effort based on this model proved successful and very promising results. The Pearson's model compares independent variables to find out possible correlations. Actually, the two sets of variables being compared are: $r_i = (x_1, x_2, ..., x_k)$, depicting the evaluations of the Learner c_i for a set $L_c = (l_1, l_2, ..., l_k)$, of commonly evaluated LRs, $k \subset m$, (m is the number of all the LRs), and: $r_j = (y_1, y_2, ..., y_k)$, which are the evaluations of the Learner c_j, for the same set of LRs.

According to the Pearson's model, the similarity $sim(r_i, r_j)$ of the two learners is given by formula (3) in Box 4, where p is the number of LRs evaluated by both Learners c_i, c_j; $x_{l,i}$ is the evaluation of c_i for the learning resource l_l; $\overline{x_{l,i}}$ is the mean value of the evaluations of c_i, from all the ranked LRs; and w_i is the weight of the Learners calculated as above from the vector of his properties, and similarly, the values that concern c_j

As in the previous case, the threshold here also depends on the data set, and is computed by the formula (4):

$$threshold_r = 1 - \frac{\sum_{j=i+1}^{n} sim(r_i, r_j)}{n * (n-1)} \qquad (4)$$

From the learners' profile, the set L_i, is derived, consisting of the LRs the Learner c_i has used, along with the action s/he has performed on them, namely, visit / download / ranking. To apply the Pearson's method, only the LRs that have been attached to c_i under the label 'rank' are used. Using these evaluations, a table nXm is created, where n is the number of Learners and m is the number of the LRs of the system. A value of 0 in a cell of this table depicts that the learner gave no evaluation for the specific item; otherwise the value of the actual rank is used.

Let's mention here, that in such a system a set of commonly evaluated LRs must exist. If no such a common ground exists no correlation can be derived between the users. Moreover, the number of these LRs must be comparable to the total number of the available LRs. When a very small number of LRs is evaluated then the formula (4) cannot conclude about their similarity. Consequently, the implementation of this algorithm is based on the hypothesis that a predefined set of minimum LRs ranked by all the learners there must exist. For a learning procedure, this means that the learner has to participate/fulfill a minimum list of mandatory activities or else that there exists a minimum amount of knowledge he has to acquire to complete the course successfully.

Consequently, let L_c the set of these inevitably evaluated LRs. For each element of L_c set each learner may provide only one evaluation.

To accelerate the algorithm, a preparation phase can precede: an ordered vector 1Xm containing 1s in the place of each mandatory LR and a 0 in the place of a non-mandatory one is used and compared to each line of the table nXm. This comparison finds out if the learner has evaluated at least the mandatory LRs. If he has not evaluated the LRs defined by this vector (the mask) s/he is excluded from the list of Learners that can be profited from the personalization utilities of

Box 4.

$$sim(r_i, r_j) = \frac{\sum_{l=1}^{p} (w_i * x_{l,i} - \overline{x_{j,i}}) * (w_j * y_{l,j} - \overline{y_{l,j}})}{\sqrt{\sum_{l=1}^{p} (w_i * y_{l,i} - \overline{y_{l,i}})^2} * \sqrt{\sum_{i=1}^{p} (w_j * y_{l,j} - \overline{y_{l,j}})^2}} \leq threshold_r \qquad (3)$$

the system. After that the formula Pearson (3) is applied for the rest of the Learners.

According to the above mentioned algorithms the members of the two neighborhoods a learner may belong, probably are different.

From each of these neighborhoods two kinds of sets of LRs derive: **a.i)** a set that contain the union of all LRs the neighbors evaluated directly; **a.ii)** a set that contain the intersection of all LRs the neighbors evaluated directly and similarly; **b.i)** a set that contain the union of all LRs the neighbors evaluated indirectly; **b.ii)** a set that contain the intersection of all LRs the neighbors evaluated indirectly. For each of these learners the sets of the LRs they have joined to their profile through direct and indirect evaluations are also known.

In the next section the functional architecture of a system is presented which applies all the above collaborative filtering methods and predicts items for a learner using a hybrid m.

4. THE ARCHITECTURE AND FUNCTIONALITY OF THE PROPOSED HYBRID CF-SYSTEM

In the core of the proposed CF-system two basic modules can be found: the **user clustering module** and the **items inference module**. The first one collects the properties and the evaluations of each learner and creates neighborhoods. The inference module finds out the set of LRs that can be retrieved for each learner.

This architecture is shown in Figure 1 and the description of the description of its functionality follows.

Learner clustering module: This module clusters the learners based on the evaluations they provided or the features of their profile. The clustering assigns the learners into groups (neighborhoods) and provides the trends of their preferences/performances. The functionality of this module is diversified based on the two following algorithms:

Clustering based on properties:

- **Step 1:** Arithmetic values are assigned to the values of each property of the learners,
- **Step 2:** The distance between pairs of learners is computed using formula (1), For n learners, the formula is applied (n-1)*n/2 times.
- **Step 3:** The average mean of the results deriving from the formula (1) is used to compute the similarity threshold (formula 2),
- **Step 4:** A learner c_i, randomly selected, is attached in the first neighborhood. Each

Figure 1. Clustering based on learner properties

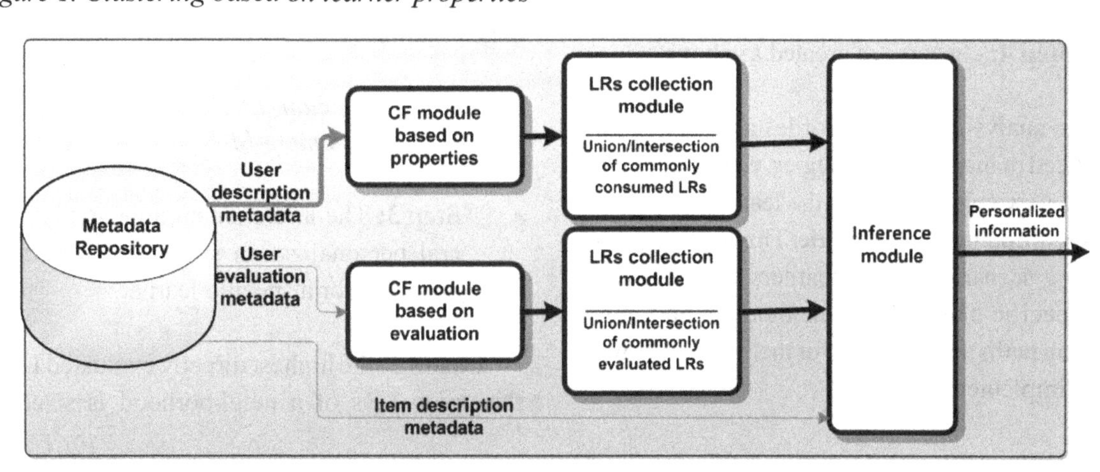

other learner $c_i\, i \neq j$ is compared to the learner c_i using formula (2). If the result of formula (2) is less than the threshold (formula 5)

$$sim(c_i, c_j) <= threshold_p \quad (5)$$

then the learner c_j belongs in the same neighborhood to c_i.

As new learners are added to a neighborhood the formula (5) must hold for the newly entered learner to each of the learners already attached to this neighborhood. This means that the formula (5) holds for all the pairs of learners that are already members of the same neighborhood.

Clustering Based on Explicit Evaluations

The second method of clustering exploits explicit evaluations. To apply this method, a minimum core of LRs evaluated on common ground by all learners should exist. To implement this method the next steps are followed:

- **Step 1:** evaluations are gathered for a set of commonly evaluated LRs,
- **Step 2:**formula (3) is applied to evaluations from pairs of learners in order to find out their relevance,
- **Step 3:** the threshold (formula 4) is computed, as in the previous method,
- **Step 4:** clusters are created as above

The analysis of these profiles anticipates for advanced management strategies, e.g., to conceptualize new evidence about the learners.

Recommendation Module: This module finds learning scenarios for the learners. The learners of a specific cluster feed this module, and new learning paths are proposed for them. The following is implemented:

- **Step 1:** for each neighborhood r the union of the LRs that have been directed ranked by all its learners is calculated: $LRtotal_r = \cup LR_i$, where r is the neighborhood of the learner ci, along with the intersection of the same LRs,

$$LRcommon_r = \cap LR_i, \textit{from each } c_i \in r,$$

  and, similarly for the LRs that have been indirectly ranked by all its learners, the union $LItotal_r = \cup LI_i$, of these LRs is computed for the neighborhood r, as well as and the intersection, $LIcommon_r = \cap LI_i$, of the these latter LRs. Of course, for each learner c_i, holds that $L_i = LR_i \cup LI_i$.
- **Step 2:** consequently, the difference between the two kind of sets are computed, the first one as derived from the union/intersection of the directly ranked LRs that the rest of the learners have attained, and the second containing the LRs that the learner c_i has already attached to his/her profile for the same reasons:

$$LU_{dreci} = LRtotal_r\text{-}LR_i,$$
$$L \cap_{dreci} = LRcommon_r\text{-}LR_i,$$

where r is the cluster the c_i belongs to, LR_{totalr} is the union of the LRs the learners of the r cluster have attached to their profile and LR_i are the LRs that c_i has attached to his/her profile. Similarly, for two sets of indirect LRs are created for each learner:

$$LU_{idreci} = LIcommon_r\text{-}LI_i,$$
$$L \cap_{idreci} = LIcommon_r\text{-}LI_i$$

- **Step 3:** The inference module applies several personalization strategies to find out elements for a specific learner.

Case 3.1: the highest directly evaluated LR of the union LRs of a neighborhood is selected:

$Ll_{\text{Recommended}} = \max(L \cap_{direcr})$, and presented to the learner, as the best recommended item.

Case 3.2: the most selected LR from the set of the $L \cap_{direcr}$ is proposed.

Case 3.3: the most selected LR from the set of the LU_{direcr} is proposed.

Case 3.4: the most selected LR from the set of the $LU_{idirecr}$ is proposed.

DISCUSSION AND CONCLUSION

Let us conclude with some notes: **i)** if the intersection of the directly evaluated LRs of the learner's neighborhood is used then it more likely to provide more precise predictions for a specific learner, but less choices are available, **ii)** if the union of indirectly evaluated LRs is used then a wider set of LRs is available, but these LRs perhaps are not the best choices align with the learner's actual profile. While items are selected from a larger group (such as the one that derives from the intersection of the directly evaluated LRs) or items are selected from a set of a pool of loosely-connected items (such as those of the intersection and union of indirectly evaluated LRs) their predictions are very likely to fail. Consequently, systems like the one proposed, which, in general, base their operation in assumptions need an indicator to measure the success of their functionality and a method to adjust/improve their outcomes. The feedback is a means to improve their performance.

Automatic System Adjustment

To find out the performance of the system the average mean $f(e_{k,\,time_1}) = \overline{er_k}$ of all evaluations that collected during a time interval, starting from initial time$_0$ up to time$_1$ is compared to the average mean of the evaluations collected from time$_1$ and up to time$_2$ using formula (6):

$$f(e) = \frac{f(e_{k,\,time1}) - f(e_{k,\,time2})}{f(e_{k,\,time1})}, \quad time_2 > time_1 \tag{6}$$

which calculates the performance of their prediction policy being applied between these two time intervals. If the result of this formula is greater than 0 then the selection of the items the algorithm made was successful, otherwise it has to be improved. In the latter case, several correction synergies can be done. E.g. **i)** a first step is to make the threshold smaller the distance between the neighbors become smaller, and they are much more similar; **ii)** a second chance is to change the set of the commonly evaluated items, so that it better represents the learners; **iii)** the selection method can be tuned as well by expanding or narrowing the set of proposed items. Exploiting further the learner's feedback new re-arrangement of neighborhoods can be achieved; **iv)** implementation of neural network techniques using the already accumulated volume if data so that we can predict the initial centre of the neighborhoods. This provides the model with the opportunity to be more accurate in future classification decisions and recommendation steps.

Future work involves estimation of accuracy of this model, and evaluations with users of a new collaborative system that detects and remediates behaviors as time spent per learning object negative attitudes towards types of items, automatic detection of typical behaviors on which the users' use the e-learning system was defined. The implementation of a Bayesian predictive distribution model, as the ones proposed by Castro (2005) and Ueno (2003), that detect irregular users' behavior and exclude them from the neighborhood creation process –after that ensuring the neighborhood's accuracy.

As a final note, the union of the LRs can be used in the beginning of the operation of a CF-based system, when a system like that usually suffers from the 'cold start' phenomenon. After a number of historical data will be gathered, the intersection can be used instead so that to make from sharp predictions.

REFERENCES

Balabanovic, M., & Shoham, Y. (1997). Content-based, collaborative recommendation. *Communications of the ACM, 40*(3), 66–72. doi:10.1145/245108.245124

Castro, F., Vellido, A., Nebot, A., & Mugica, F. (2005). Applying data mining techniques to e-learning problems. [SCI]. *Studies in Computational Intelligence, 62*, 183–221. doi:10.1007/978-3-540-71974-8_8

Herlocker, J.-L., Konstan, J.-A., & Riedl, J. (2000). Explaining collaborative filtering recommendations. *Proceedings of the 2000 ACM conference on Computer supported cooperative work*, Philadelphia, Pennsylvania, United States, (pp. 241-250).

Itmazi, J., & Megias, M. (2008). Using recommendation systems in course management systems to recommend learning objects. *The International Arab Journal of Information Technology, 5*(3), 234–240.

Kerkiri, T., Konetas, D., Paleologou, A., & Mavridis, I. (in press). Semantic Web technologies anchored in learning styles as catalysts towards personalizing the learning process. *International Journal of Learning and Intellectual Capital.*

Kerkiri, T., Mavridis, I., & Manitsaris, A. (2009). How e-learning systems may benefit from ontologies and recommendation methods to efficiently personalize knowledge. [IJKL]. *International Journal of Knowledge and Learning, 5*(3/4), 347–370. doi:10.1504/IJKL.2009.031229

Kerkiri, T., Paleologou, A., Konetas, D., & Hatzinikolaou, K. (2010). A learning style–driven architecture build on open source LMS's infrastructure for creation of psycho-pedagogically–savvy personalized learning paths. In Soomro, S. (Ed.), *E-learning experiences and future*.

Kritikou, Y., Demestichas, P., Adamopoulou, E., Demestichas, K., Theologou, M., & Paradia, M. (2008). User profile modeling in the context of Web-based learning management systems. *Science Direct. Journal of Network and Computer Applications, 31*(4), 603–627. doi:10.1016/j.jnca.2007.11.006

Parra, D., & Brusilovsky, P. (2009). Collaborative filtering for social tagging systems: An experiment with CiteULike. Proceedings of the Third ACM Conference on Recommender Systems, New York, USA, (pp. 237-240).

Resnick, P., & Varian, H. (1997). Recommender systems, introduction to special section. *Communications of the ACM, 40*(3).

Resnick, P., Zeckhauser, R., Friedman, E., & Kuwabara, K. (2000). Reputation systems: Facilitating trust in Internet interactions. *Communications of the ACM, 43*(12), 45–48. doi:10.1145/355112.355122

Salton, G., Wong, A., & Yang, C.-S. (1975). A vector space model for automatic indexing. *Communications of the ACM, 18*(11), 613–620. doi:10.1145/361219.361220

Schafer, J., Frankowski, D., Herlocker, J., & Sen, S. (2007). Collaborative filtering recommender systems. In Brusilovsky, P., Kobsa, A., & Nejdl, W. (Eds.), *The adaptive Web* (pp. 291–324). Berlin/ Heidelberg, Germany: Springer-Verlag. doi:10.1007/978-3-540-72079-9_9

Tsai, K.-H., Chiu, T. K., Ming, C.-L., & Wang, T.-I. (2006). A learning objects recommendation model based on the preference and ontological approaches. *Proceedings of the Sixth International Conference on Advanced Learning Technologies* (ICALT'06), (pp. 36-40).

Ueno, M. (2003). *LMS with irregular learning processes detection system.* In World Conference on e-Learning in Corp., Govt., Health, & Higher Education. (pp. 2486-2493).

Upendra, S., & Patti, M. (1995). Social information filtering: Algorithms for automating word of mouth. *Proceedings of the SIGCHI conference on Human factors in computing systems,* (pp. 210-217).

Wang, T. I., Tsai, K. H., Lee, M. C., & Chiu, T. K. (2007). Personalized learning objects recommendation based on the semantic-aware discovery and the learner preference pattern. *Journal of Educational Technology & Society, 10*(3), 84–105.

Wikipedia. (2010). *Collaborative filtering.* Retrieved from http://en.wikipedia.org/wiki/Collaborative_ filtering

Wikipedia. (2010). *Reputation system.* Retrieved from http://en.wikipedia.org/wiki/Reputation_ system

Wikipedia. (2010). *Social software.* Retrieved from http://en.wikipedia.org/wiki/Social_software

Ziegler, C. (2004). Semantic Web recommender systems. In Lindner, W. (Ed.), *EDBT 2004Workshops* (pp. 78–89). Berlin/Heidelberg, Germany: Springer-Verlag.

Chapter 13
Student Models Review and Distance Education

Avgoustos A. Tsinakos
University of Kavala Institute of Technology, Greece

ABSTRACT

The current chapter is a review of the variety of student models that have been reported in the literature. The chapter can virtually be divided in two parts: The first part outlines a number of typical examples of student models that have been developed in order to indicate why these models have been developed, what are their uses and the achievements of student models in education. The second part discusses how the student models can be useful in distance education, what are the criteria for testing the applicability of such models and finally reports student models that may apply in asynchronous distance education.

WHAT ARE STUDENT MODELS?

Definitions. "In general terms, Student Modelling involves the construction of a qualitative representation that accounts for student behavior in terms of existing background knowledge about a domain and about students learning the domain. Such a representation is called a Student Model" (McCalla, 1992 a).

Other Student Model related definitions come from Barr and Greer. In more detail:

DOI: 10.4018/978-1-61520-921-7.ch013

"Student Model represents student understanding of the material to be taught with the purpose to make hypotheses about student's misconceptions and suboptimal performance strategies" (Barr et al, 1982)

Jim Greer illustrates in bullets the different interpretations that a Student Model might have indicating that such model may be:

- An abstract representation of the learner.
- Teacher's conceptualisation of a learner.
- System's beliefs about the learner.

- System's beliefs about the learner's beliefs and skills.
- It may include history of learner actions (raw data)
- Interpretations of raw data.
- Explanations of behaviour (Greer, 1996).

He also indicates that "Student modelling is the process of acquiring knowledge about a learner, relative to the learning goals, and although some a priori knowledge might have been acquired about this learner or similar learners, normally the knowledge acquisition is carried out dynamically as the learner engages in interactions with the system" (Greer, 1996).

Types of Student Models

According Anderson, Corbett and Koedinger there are actually two types of student modelling the *knowledge tracing* and *model tracing* (Anderson *et al.*, 1995, p. 167–207). These are their names refer to the particular techniques they used, however the distinction is somehow general. Knowledge tracing refers to the problem of determining what students know, including both correct domain knowledge and robust misconceptions.

Model tracing refers to tracking a student's problem solving as the student works on a problem. Model tracing is useful for systems that attempt to answer requests for help or to give unsolicited hints and feedback in the middle of problem solving. In fact, to be adequate in assisting, hinting and assessing an on-going solution attempt a system has to understand at a minimum level what line of reasoning the student is attempting to pursue.

On the other hand, knowledge tracing is useful for making longer range pedagogical decisions, such as what problem to assign next or what evaluation grade to assign to the student.

Giangrandi Paolo and Carlo Tasso, who have criticised the concept of static and of temporal Student Models, provide an additional differentiation on the concept of student modelling. According to their definition a model which describes the student's knowledge without considering the possible evolution in time is called *static*. A description of the temporal history of the student's knowledge including all the information about the student which makes it possible to explain the student's behavior for both the past and the current interaction is called *temporal*. (Giangrandi and Tasso, 1996, p. 184-190; Giangrandi and Tasso, 1997, p. 415-426).

MaCalla introduces two other types of Student Models, the explicit and the implicit, and makes a useful distinction between them (McCalla, 1992 b, p. 107-131). An explicit Student Model is a representation of the learner in the learning system that is used to derive instructional decisions. An implicit Student Model is reflected in design decisions that have been derived from the system designer's view of the learner.

Core Characteristics/Components of a Student Model

Clancey, Self, McCalla and Greer introduce some of the core characteristics of Student Models.

According to Clancey, Student Models are qualitative models in the sense that they are neither numeric nor physical; rather, they describe objects and processes in terms of spatial, temporal, or causal relations (Clancey, 1986, p. 381-450).

On the other hand according to Self, Student Models are approximate, possibly partial, and do not have to fully account for all aspects of student behaviour they are interested in computational utility rather than in cognitive fidelity. A more accurate or complete Student Model is not necessarily better, since the computational effort needed to improve accuracy or completeness may not be justified by the extra if slight pedagogical leverage obtained (Self, 1994).

McCalla and Greer, described the kind of architecture they conceive for a Student Model. According to them, the tutor's knowledge is divided into a number of conceptual spaces such as: the tutors goal space, the tutor's belief space, the student's goal space, and the student's belief space (Figure 1).

The tutor's goal space includes the pedagogical goals of the tutor and pedagogical planning algorithms to turn these goals into instructional plans.

The tutor's belief space includes the knowledge the tutor has about the domain of concepts, which the student is exploring. It includes knowledge about tasks, which illustrated these concepts, as well as knowledge about common misconceptions. These two spaces allow the tutor to explicitly reason about his/her own knowledge and pedagogical goals.

The student's goal space and the student's belief space represents the tutor's beliefs about the student and together constitute the Student Model. The student's goal space contain the tutor's perceptions of the cognitive goals of the student as well as plan recognition algorithms, which the tutor can use to induce a student's plans for learning the concepts in the tutoring domain.

The student's belief space contains tutor's perceptions of what domain-related information the student understood. The student's belief space has two interesting subspaces: the student's beliefs about the tutor's beliefs and the student's beliefs about the tutor's goals. These subspaces are essential if the tutor is to predict the student's reaction to the tutor (McCalla and Greer, 1994, p. 39-62).

Raymund and Masamichi, indicate three essential elements of student modeling: the student behaviour, the background knowledge and the Student Model.

Student behaviour refers to a student's observable response to a particular stimulus in a given domain, which, together with the stimulus, serve as the primary input to a student modeling system. This input (i.e., the student behaviour) can be an action (e.g., writing a program) or, more commonly, the result of that action (e.g., the written program).

Background knowledge is comprised of the correct facts, procedures, concepts, principles, schemata and/or strategies of a domain -called theory- and of the misconceptions held and other errors made by a population of students in the same domain –called bug library.

Student Model is an approximate, possibly partial, primarily qualitative representation of student knowledge about a particular domain, or a particular topic or skill in that domain, that can fully or partially account for specific aspects of student behaviour (Raymund and Masamichi 1998, p. 128-158).

Student Model vs. User Model

A critical issue of the study was the notification of the difference between the user model and the

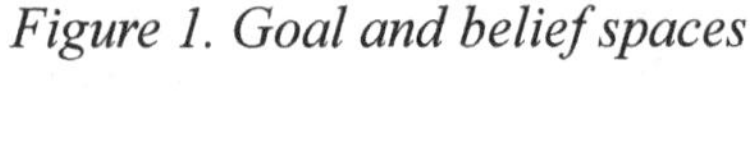
Figure 1. Goal and belief spaces

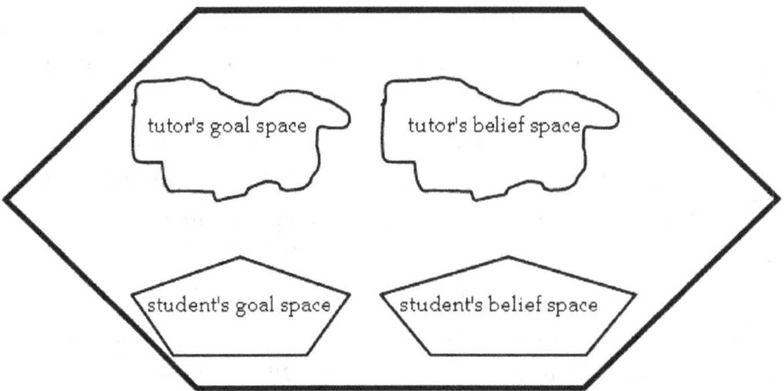

learner or Student Model. According Greer and McCalla

The process of User Modelling deals with modelling users of computer systems that support on-line coaching and consulting. Although it may address similar issues to those addressed by Student Modelling such as knowledge representation, plan recognition, diagnosis etc, User Modelling largely ignores learning theory and the fact that the type of interaction with the user can be quite different from pedagogical interactions (Greer and McCalla, 1994, p. 26).

Martina Auberger had also reported another difference:

The user model as used in human computer interaction (HCI) is generally external to the application and is not explicitly stored or coded within the application. The student model in computer based education (CBE) in its fully developed form (as an intelligent tutoring system - ITS) is explicitly stored or coded within the application. Computer based education is the support of education by a computer in order to facilitate the learning process. However, many solutions and ideas used in either of the models could also be used by the other one. In the most general sense, a student model is the system's beliefs about the learner. Along with his knowledge this comprises both the capabilities of the learner and his learning preferences. Other properties are his individual way of learning, his misconceptions about a domain, etc. (Auberger, 1998, p.5)

USES AND ACHIEVEMENTS OF STUDENT MODELS IN EDUCATION

Student modeling is defined as the task of describing the knowledge and beliefs of the student as a basis for the decision on appropriate actions for feedback. The subject of how the students learn and acquire the knowledge gained major interest by the teachers in conventional education. Over the years teachers have changed their strategies because the knowledge of how students learn increased. These strategies were subject to change as teachers kept on learning about students. Development of Artificial Intelligence and of Multimedia provided new horizons for the evolution of Student Models.

Student Models had been extensively used in the development of:

1. Intelligent Learning Environments,
2. Intelligent Tutoring Systems,
3. Intelligent Computer Aided Instruction Systems,
4. Multimedia Educational Environments
5. and lately of Web based Instruction Systems.

Extensive use of Student Models derived from a number of different reasons. One of the basic intentions was to monitor and understand student reasoning and misconceptions- a quite difficult task -since student reasoning was not directly observable. The identification of ways to adapt and guide teaching on individual basis according student's cognitive characteristics and learning behaviour was a different approach. Yet another reason was the purpose of performance assessment and certification of student's mastery on a given context.

Several different uses of Student Models had been reported in the literature (VanLehn, 1988; Nwana, 1991; Holt et al. 1994). Eva Ragnemalm (1999) has identified that the uses of Student Models can be divided in four categories

1. **Planning education:** What topics are to be learned? Which are well known? Choosing exercises at the edges of the student's capacity requires knowing what is known and what is not known in a general sense
2. **Planning delivery:** What experiences are suitable to encourage learning of the intended topic and which previous experiences can be utilized? Assistance such as providing hints during the problem solving is based on

known subjects. Sometimes new concepts must be introduced, sometimes they can be discovered or built on old ones already known requiring detailed understanding of previous knowledge.

3. **Generating feedback:** Feedback on performance should build on previous knowledge as well as current conditions. Feedback concerning well knowing subjects is different from feedback on subjects recently learned.
4. **Remediating misconceptions:** Remediating misconceptions can be done by pointing them out to the student, either by providing counter evidence or by heaving a meta-level discussion. (Ragnemalm, 1999, p. 17)

Yet another aspect that motivated the use of Student Models was the recording of students' learning style (by examining their learning speed) and their motivation to learn (by encouraging collaboration or competition with peer students).

The general achievements gained by the use of Student Models in education were the inspection and analysis of students' mental behaviour, of their reasoning and of the knowledge that was believed to underlie such behaviour. In the following list some particular achievements are reported as they appeared in the literature resulted from the use of Student Models:

1. Elicitation of students' misconception from observed errors during a problem solving process (Matsuda & Okamoto, 1992).
2. Diagnosis of students behavior in a procedural problem solving session. (Matsuda & Okamoto, 1994)
3. Assessment of students' expertise on a given domain (VanLehn & Martin1997, p. 179-221; Conati & VanLehn, 1996; Katz & Lesgold, 1992, p. 205-230).
4. Construction of bug libraries based on students' misconceptions (Greer & McCalla, 1994, p. 8; Baffes & Mooney, 1996).
5. Long-term knowledge assessment, plan recognition and prediction of students' actions during problem solving (Conati *et al.,* 1997, p. 231-242).
6. Inference of students' problem solving ability, acquisition of new topics and retention of earlier topics (Beck & Park & Woolf, 1997, p. 277-288).
7. Representation of the granularity of knowledge (Collins et all.,1996) (Eva pg 23)
8. Improvement of students' performance by provoking either student-teacher interaction or interaction with a peer student (Bull, 1997; Bull & Brna, 1999).
9. Support of students' writing skills (Bull & Shurville, 1999)
10. Assessment of student's self explanation (Conati & VanLehn, 1999).

TYPICAL EXAMPLES OF STUDENT MODELS THAT HAVE BEEN DEVELOPED

A number of Student Models have been reported in the literature. Some of the implemented models were domain independent while some others had domain dependency.

Domain Independent Student Models

Jim Greer and Gordon McCalla, in their book "Student Modelling: The Key to Individualized Knowledge-Based Instruction" (Greer and McCalla, 1994) report a number of domain independent Student Models introduced by Kass (Kass, 1989, p. 386-410).

Overlay Student Model

According to this model the learner's knowledge at any point is considered as a subset of an expert's knowledge. The objective of instruction is to establish the closest possible correspondence between the two sets. In such approach it is assumed that all differences between the learner's behaviour and that of the expert model can be explained as the learner's lack of skill.

The Overlay Student Model works well for systems where the goal is to strictly impart the knowledge of the expert to the learner. On the other hand the main problem with this model is that the model assumes that a learner's knowledge can be merely a subset of an expert which may not always be the case (Greer and McCalla, 1994, p. 7)

Differential Student Model

According to the Differential Student Model the learner's knowledge is divided into two categories: knowledge that the learner must know and knowledge that the learner can not be expected to know (Figure 3)

So in contrast with the overlay Student Model this model does not assume that all gaps in the Student Model are equally undesirable. Furthermore the differential Student Model acknowledges and tries to represent explicitly both learner knowledge and learner-expert differences.

Perturbation Model and Bug Library

While the overlay model represented the learner only in terms of "correct" knowledge, a perturbation model normally combines the standard overlay model with a representation of faulty knowledge (Greer and McCalla, 1994, p. 8). Furthermore, in the perturbation model the learner is not considered as a mere subset of the expert, but the learner possesses knowledge potentially different in quantity and in quality from the expert knowledge.

The perturbation model maintains a close link between the learner and the expert concepts but also represents the learner's knowledge and beliefs beyond the range of the expert knowledge (Figure 4).

This model utilises a bug library, which is a fixed collection of bugs and students' misconceptions. The inclusion of the bugs in the perturbation model allows more sophisticated understanding of the learner to be accomplished than with a simple overlay on the expert model.

Constraint-Based Student Model

A constraint-based Student Model represents the learner knowledge as the constraints upon the

Figure 2. Overlay student model

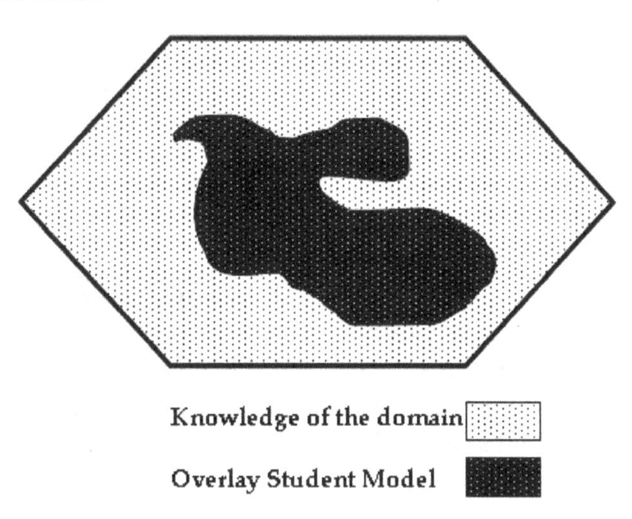

Figure 3. Differential student model

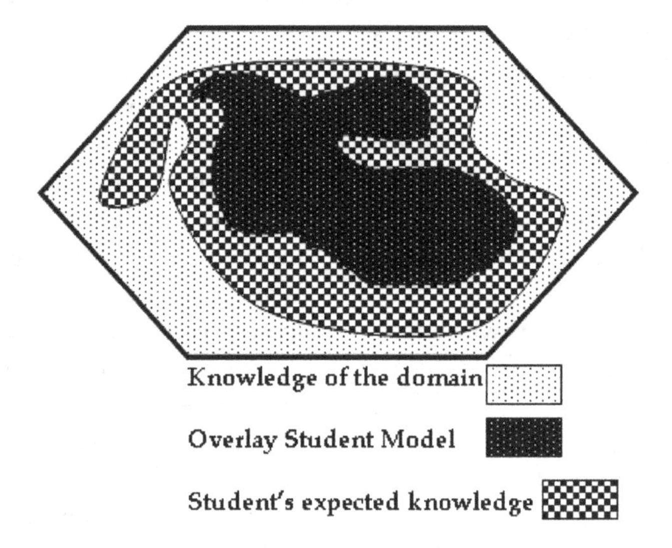

Figure 4. Perturbation model

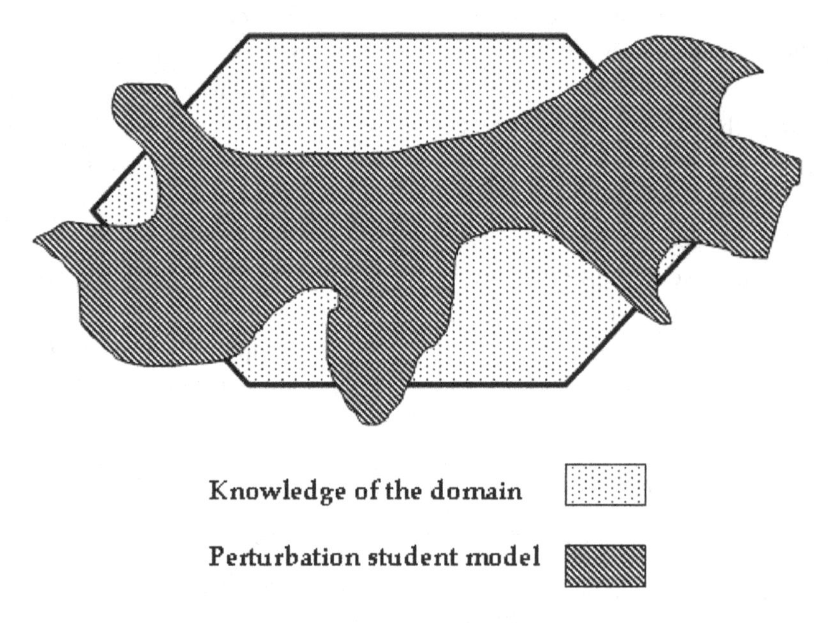

correct knowledge representation. It extends the standard overlay model approach by permitting much more sophisticated reasoning about domain concepts beyond whether, they simply knowing them or not A violation of those constraints by the learner indicates that the model needs to be updated. This model is characterised by computational simplicity and does not prescribe a particular tutorial strategy. Furthermore it is unclear how this approach would generalise across domains and tutorial strategies.

PairSM Student Model

PairSM is a domain-independent system introduced by Susan Bull and Matt Smith aimed at helping pairs of students to organise their revision for an approaching test (Bull and Smith, 1997, p.

Figure 5. The two student models of pairSM

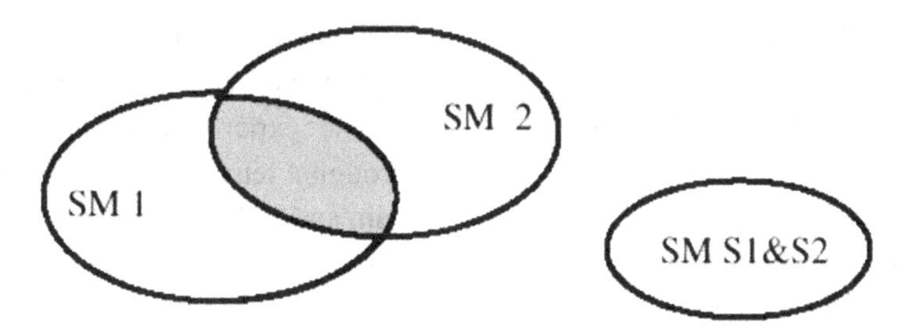

339-341). PairSM contains two individual Student Models that are compared by the system to enable it to suggest ways in which two students may work together effectively. It recommends collaborative learning, peer tutoring or individual learning, depending on the comparative contents of the models. The aim is to encourage students to experience the benefits of peer interaction.

The two Student Models of PairSM are initially based on the results of a multiple choice pre-test entered by the tutor. The models are updated by subsequent tests. PairSM contain heuristics for recommending the kind of preparation which might be useful for the learners by comparing the contents of the two Student Models and the manner in which the learners have acquired further knowledge. An overview of the Student Models is given in Figure 5.

The concepts known by Student 1 were represented in SM 1 while those known by Student 2 in SM 2. The intersection of SM 1 and SM 2 represents the shared knowledge. SM S1&S2 represent knowledge that the two students can produce when working together, but not individually. (Bull and Smith 1997, p. 339-341).

PeerISM Student Model

PeerISM is another domain-independent system introduced by Susan Bull and Paul Brna. It intends to be used with any domain which has potential for profitable peer interaction.

PeerISM is an inspectable Student Model aimed to promote effective peer interaction enabling the students to built their own model of a domain which may be improved through reciprocal student modeling with a peer student and subsequent face-to face peer negotiation of a domain model (Bull & Brna, 1999).

In peerISM the students are required to assess their responses to the questions posed by the tutor, to provide feedback to a peer students who answered the same questions, and to negotiate a model of the domain in cases where the self assessment and the peer evaluations are incompatible.

As reported in the paper:

Self and peer assessments have both qualitative and quantitative components. The quantitative evaluations from both sources comprise the explicit representations of the student models in the peerISM system. In providing a self evaluation, learners are constructing a simple model of their confidence in a particular area, for peerISM. When giving feedback, they are creating a simple model of their partner's beliefs which may or may not coincide with their own beliefs. (Bull & Brna, 1999)

Beyond students' comments and feedback, feedback is also provided by the peerISM system in the form of qualitative comments indicating misconceptions that occurred or the adequate performance of the students.

STUDENT MODELS WITH DOMAIN DEPENDENCY

See Yourself Write Student Model

The See Yourself Write is a system comprising of two parts: a template through which the teacher provides the feedback on each writing assignment and an individual Student Model which is automatically created from the feedback provided by the teacher which is built up over time (Bull, 1997, p. 315-326). The Student Model of *See Yourself Write* is inspectable, to promote learner reflection and to encourage learners to use the feedback received from their tutors in their future assignments.

The domain that this model was applied was writing in a foreign language.

See Yourself Write differs from other computational writing environments containing a learner model because the aim is differentiated. The aim of the *See Yourself Write* Student Model differs from that of more conventional learner models in that it is not intended as a source of information for a computational educational system, but rather as a source of information *for the student*. It reflects to students' feedback on their own work, and information about how they are progressing and about their overall performance. Information in the Student Model is both qualitative and quantitative. Its main purpose is to promote learner reflection on completed assignments in such a way as to lead students to use this feedback to improve subsequent work.

Since the *See Yourself Write* model is constructed from teacher feedback, the writer is not constrained in anyway during the writing session. The inferencing which took place in the construction of the model, occurred away from the learner. Because the Student Model aims to help students think about and take control of their learning, the qualitative textual descriptions produced by the teacher are also an important component of the model.

The term *Student Model* was therefore broader in this context than was customary, including both system evaluation and teacher feedback.

The expert teacher provides the feedback through a template created for the particular domain and task requirements (writing in a foreign language).

Once the feedback is entered into the template to the teacher's satisfaction, the teacher sent it to the separate Student Model, where qualitative information is placed in the appropriate areas for student retrieval, and quantitative information is similarly distributed for ease of viewing. In addition, the quantitative feedback is evaluated by the system to provide a more general overview Student Model to be used by the student in conjunction with feedback on individual assignments.

HOW CAN STUDENT MODELS BE USEFUL IN DISTANCE EDUCATION?

The development of new technologies has promoted an astounding growth in distance education, both in the number of students enrolling and in the number of universities adding education at a distance to their curriculum. A great number of students of different ages participated in courses available via distance.

Teaching and learning at a distance required special attention to the possibilities and limitations of the technologies involved. Even many simple activities, which were taken for granted as part of the instructional process had to be reconsidered and adapted to new circumstances. Pedagogy of distance education had to be broken in issues that concerned the instructor, and issues that concerned the student. The area of Distance Education can be positively affected by the use of Student Models promoting the achievement of better teaching and learning activities.

Student models can be used in an Asynchronous Distance Education in a similar way to the one that has already been used in traditional education. As it was earlier reported Student Models are used to elicit information from the students regarding their misconceptions or their weaknesses during the learning session or regarding their learning pace. Furthermore Student Models are used to provide help and feedback to the students according to their learning preferences. Such beneficial use of Student Models can also be valid in Asynchronous Distance Education. However, a critical difference appears when comparing the use of Student Models in Asynchronous Distance Education to the ones used in traditional education.

In the latter case Student Models are used as one of the central components of Intelligent Tutoring Systems or of Multimedia Educational Environments. The main scope in traditional education is the achievement of an improved simulation of a virtual tutor or of a peer student so that more effective educational sessions are achieved when students interacted with the system. So the central idea is the improvement of individualised learning as the system-guided by the Student Model component- tried to perform the best adaptation to the learning preferences of each individual student. Thus, Student Models are considered as the backup support premise of the computer based educational systems in order to meet students' learning needs. In most cases Student Models are not observable by the students and they are used for internal computational implementations.

The above situation is differentiated when confronted with an Asynchronous Distance Education session. Although the main idea of supporting students' learning needs remained valid the use of Student Models has a totally different orientation. The key concept in this case is not the improvement of the adoption of some computer based educational environment but how to help both the students and the teacher to achieve more effective educational sessions.

The learning material in Asynchronous Distance Education can be accessed or read by the students at any time. Similarly, questions can be asked asynchronously to other classmates as even the study pace can also be different. This discontinuity of the education session in addition to the lack of face to face contact causes a pedagogical disadvantage affecting both the tutor and the students.

Tutors may become confused or incapable to monitor the individual learning needs or the misconceptions of the distance students. Tutors may also have problems to locate a particular e-mail among the others in their m-boxes, which addressed a study problem of a student. The same problem may be extended to the inability of the tutors to keep a track of the feedback provided to each student, especially when a large group of students is supervised.

Students on their side may felt isolated being at "a real distance" from their tutor and their classmates. Lack of intense supervision related to students' individual learning needs may also be recognized in addition to the lack of effective support or guidance on how to overcome a misconception or remedy a performance gap.

Thus Student Models used in Asynchronous Distance Education may act as an intermediate communication component between the tutor and the students recording tutor's suggestions and feedback regarding students' progress in addition to the comments made by the students regarding their personal problems or the misconceptions occurred.

Criteria Testing the Applicability of Student Models in Distance Education

The criteria of testing the applicability of Student Models in Asynchronous Distance Education can be categorized as follows:

1. Generic criteria, which concern the general features of a Student Model such as implementation issues, domain dependency etc.
2. Specific criteria, which concern specific pedagogical issues that the Student Model

should support such as the promotion of collaboration, improvement of context comprehension etc.

In more detail the generic criteria used to examine the applicability of the Student Models are:

1. *They should accurately explain student behaviour independently of the domain tackled.* Due to the variety of the educational subjects provided via distance, the Student Models should be able to capture and explain students' behaviour regardless of the context of the taught course. This assumption though does not imply that the entire domain dependent Student Models that have been developed are inapplicable in the area of Distance Education. However reviewing domain dependent Student Models towards each educational subject available via distance would be impractical.
2. *They should be easy to handle and to maintain.* For practical reasons Student Models should be easy to handle and to maintain by the instructor. In case of Distance Education the problem does not focus only on just collecting some statistical data for computational use regarding students' progress. On the contrary such models should assist the instructors to understand and easily monitor students' performance on a continuous basis and during the educational session.
3. *They should be inspectable by the student and available for consulting on student's request.* The context of a Student Model should not be used only for computational implementations kept hidden from the students. On the contrary the context should be kept in a meaningful -and easy to be used by the student- format, continuously accessible whenever the student requested it.

The specific criteria met by a Student Model applicable in an Asynchronous Distance Education are:

1. *Should act beneficial for both the tutor and the student.* In conjunction with the third generic criterion, the need for simplicity and ease of understanding Student Models derived from the fact that Distance Education is addressed to students with great variety of educational background. The majority of the student population is not aware of or familiar with the use of complicated software products. Due to the lack of physical tutor-student contact sometimes the distance student have the felling that the teacher is unreachable when needed. This is the reason why Student Models should provide bi-directional benefit to both instructors and students by enabling students to monitor their own progress and utilise the feedback signals propagated by the model on a continuous basis. In that way Student Models can be simulated to an off-line student support system by remedying student performance problems and by reporting their progress whenever the student access them.
2. *Should contribute to the improvement of context comprehension* by highlighting students' misconceptions or students' poor performance on a given topic and further recording interactions between the students and the teacher or students' peer tutoring sessions.
3. *Should promote student's self reflection and self explanation* not only by reporting students' misconceptions but also the reasons that they occurred.
4. *Should facilitate students' supervision* by enabling the tutor to have a solid and continuous view of students performance including both quantitative and qualitative information in addition to students' individual weaknesses and strengths

Report of Student Models that May Apply in Distance Education

The models evaluation has been carried out according to the above mentioned criteria and has been separated in two sections for each model; the Generic section (G) and the Specific section (S).

- **(G)** The *Overlay Student Model* is domain independent and has an easy representation of both knowledge spaces (tutor's and student's). A clear final target can be identified which is to increase student's knowledge space as much as it is possible so that the differences among the two knowledge spaces is minimised.

This model may employ objective methods in the creation of learning instructions as it expects a minimum prior knowledge and skills on student's part towards the taught domain. Objective method provide a series of steps that lead the learner to the final goal (behaviour that the student should have by the end of the instructional session i.e. the student should be able to save documents on a floppy disk).

There is no information available if such model is inspectable by the student neither on how it is updated as the learners' knowledge increases. Thus, no conclusion can be drawn on the ease of handling and maintenance of the model or its inspectability.

- **(S)** Overlay model can be beneficial for both the student and the teacher and facilitates students' supervision as the model represents the growth of the students' learning space.

On the other hand the model does not provide facilities to promote students' self reflection nor any kind of indications of improvement on context comprehension

- **(G)** The *Differential* and the *Perturbation Student Models* assume that all gaps in the Student Model are not equally undesirable, nor that the student's knowledge is considered as a mere subset of the expert (tutor). For example a distance student may not be able to indicate the characteristics of the self paced corresponding studies but to be able to develop effective and well designed web pages while the instructor can not. Such differences in student's and in tutor's knowledge space are not undesirable, neither can they rank the student as a novice. On the contrary, the learner is assumed to possess knowledge potentially different in quantity and in quality from the expert knowledge.

These models similar to the Overlay Student Model are domain independent and they represent both student knowledge and student-expert differences more explicitly. Furthermore similarly to the Overlay model no information exists on the inspectability or the maintenance of the two models

Both the Differential and the Perturbation models may employ constructivist methods in the creation of learning instructions allowing the student to take control of the learning.

- **(S)** The two models satisfy all the specific criteria with special reference to the Perturbation model where the utilisation of bug library provides information promoting students' self reflection and hints on context comprehension.
- **(G)** Similar analysis to the Overlay model was assigned to the *Constraint Based Model*. Though some variation points existed.

According to the first generic criterion this model can only be applied on deterministic domains while referring to the second criterion the model seems easy to be maintained since no computational complexity of the model exists (once constraints are violated, update of the model is required).

- **(S)** Another variation on the specific criteria is the fact that the model promotes students' self reflection as the student is represented as the constraints upon the correct knowledge representation.
- **(G)** *PairSM* is also a domain independent model that has exceptional interest as it can be used to monitor collaborative learning - a critical issue- in the area of distance education.

This model has clear representation of students' knowledge space as team and as individuals. The model is also updated on continuous basis by the subsequent tests entered by the tutor.

Although the model has an easy maintenance, no information exists on the inspectability of the constructed Student Models by the students themselves.

- **(S)** In pairSM once the students have individually completed the tests, the system is able to consider where help is needed most, and whether the students can usefully collaborate or tutor each other. Students can later return to PairSM to take another test. Their first test was taken separately, but if the students wish to take advantage of all of the possible representations for SM-S1S2 they retake the same test in a situation where they can communicate. The system is then able to compare the results of the collaboratively taken test with the identical, but individually taken version, before making its next recommendations (Bull and Smith 1997, p. 339-341).

In conclusion this model satisfies all the specific criteria as the collaborative learning contributes to the context comprehension and to students' self reflection and explanation. Furthermore the tutor via the system is able to monitor students progress both as group and as individuals.

- **(G)** *PeerISM* can be considered as one of the most appropriate models to be applied in an Asynchronous Distance Education session. The model being domain independent seems an attractive solution for Distance Education since it is easy to be maintained and handled as it is based on the analysis of students' textual interactions.

The constructed models are inspectable on both sides, the tutors' and the students'. The models can also be modified after a number of interactions among the students.

- **(S)** The model also satisfies all the specific criteria. In more detail collaboration interaction about students' knowledge aims to enhance understanding of the domain and to raise awareness of the learning process.

Self-assessment and colleague evaluation facilitate the self-reflection and explanation to a great extent. Furthermore, the model acts beneficially on students' supervision as both quantitative and qualitative information is stored in the model.

The only problem seems to exist in cases where both students share the same misconception. In such case there is no information available on how the system will identify the problem since system's indication of misconceptions occurs only when incompatibility on students-self evaluation is detected.

- **(G)** Finally the *See Yourself Write* Student Model is implemented to improve writing skills in a foreign language. It could be used as a domain independent Student Model aiming to improve student's performance by utilising tutor's feedback. This model is considered to be the most appropriate model to be applied in Asynchronous Distance Education as it successfully meets all the generic and specific evaluation criteria.

The model is easy to handle and to be maintained on teacher's side as the latter could enter his/her comments and view the results in plain text without complicated computational procedures to be necessary. Tutor's additional comments are also available to the student's eyes and the student can also argue on these comments by entering his/her point of view or keeping notes regarding the reasons that have caused a potential faulty performance. In this way a bi-directional interaction between the student and the teacher is achieved while the Student Model acts as an intermediary system.

- **(S)** The above feature also affects the specific criteria especially the improvement of context comprehension in addition to the promotion of student's self-reflection and explanation.

In most of the courses available via distance the tutor provids feedback to the students through the assignments trying to help them to improve their performance or to highlight their weak points.

This model utilises to a maximum level this process of feedback provision. The system not only keeps both quantitative and qualitative comments for each particular assignment but it also automatically constructs and updates an individual Student Model based on the feedback provided each time by the teacher via the corrected assignments.

This individual Student Model includes general qualitative comments on student's performance and the overall Student Model derives by the quantitative assessment of each assignment.

Furthermore the tutor can type additional comments on a separate template in case further explanations are required.

As Susan Bull indicates in her article *See Yourself Write* was aimed at helping learners to reflect on their performance, and to think about how they might improve their work by:

- viewing and interacting with a student model based on teacher feedback;
- making it easy for students to access useful comments on earlier work when composing a new piece of work;
- being prompted to explain how they could improve, or to give reasons to explain their improvement/deterioration, etc.;
- being encouraged to take advantage of self-explanation of difficulties by disguising this as a request for outside assistance.

Additionally the facility for students to disagree with the model also enables teachers to become aware of their mis-diagnoses (Bull, 1997, p. 315-326).

As final remark the See Yourself Write Student Model has a different aim from that of more conventional learner models; "that it was not intended as a source of information for a computational educational system, but rather as a source of information for the student". This is one of the central strength points of the See Yourself Write Student Model.

SUMMARY

The present study was designed to investigate the use and the applicability of Student Models in education and in Asynchronous Distance Education in particular. A variety of approaches has been employed for the development of Student Models including applications of the area of artificial intelligence, cognitive psychology and instructional science. Fuzzy set theory and Bayesian statistics have been applied to a number of Student Models as those introduced by Katz, Lesgold, Eggan and Gordin (Greer and McCalla, 1994, p. 31).

Some of the introduced models are domain independent and furthermore generic (such as the

Overlay and the Perturbation Student Models), while some others are domain dependent (such as the See Yourself Write). Part of the benefits gained from the use of Student Models in education are the inspection and analysis of students' mental behaviour in addition to tracking students' reasoning and underlying knowledge.

In the current study the issue of the applicability of some of the developed Student Models on the area of Distance Education was examined. It has to be mentioned that a new role was assigned to the use of Student Models in Asynchronous Distance Education. Models in this case act as an intermediate communication component between the tutor and the students, recording tutor's suggestions and feedback regarding students' progress in addition to the comments made by the students regarding their personal problems or the misconceptions occurred.

In order to evaluate the applicability of the reviewed Student Models, both generic criteria –concerning general features of the models such as domain dependency- and specific criteria- concerning pedagogical issues- have been used. The study indicated that pairSM peerISM and See Yourself Write in particular were the most appropriate models to be used in an Asynchronous Distance Education session.

REFERENCES

Anderson, J., Corbett, A., Koedinger, K., & Pelletier, R. (1995). Cognitive tutors: Lessons learned. [NJ: LEA.]. *Journal of the Learning Sciences*, *4*(2), 167–207. doi:10.1207/s15327809jls0402_2

Auberger, M. (1988). *Student modeling in educational multimedia titles using agents*. Department of Information Business, Vienna University. Retrieved from http://lux2.wu-wien.ac.at/~geyers/archive/auberger/proposal/proposal.html

Baffes, P., & Mooney, R. (1996). Refinement-based student modeling and automated bug library construction. *Journal of Artificial Intelligence in Education*, *7*(1), 75–116.

Barr, A., Beard, M., & Atkinson, R. C. (1982). The computer as a tutorial laboratory: The Stanford BIP Project. *International Journal of Man-Machine Studies*, *8*, 567–596. doi:10.1016/S0020-7373(76)80021-1

Beck, J., Stern, M., & Woolf, B. P. (1997). Using the student model to control problem difficulty. *User Modeling: Proceedings of the Sixth International Conference*, UM97 Vienna. (pp. 277-288). New York, NY: Springer Wien.

Bull, S. (1997). See yourself write: A simple student model to make students think. *User Modeling: Proceedings of the Sixth International Conference*, UM97 Vienna. (pp. 315-326). New York, NY: Springer Wien.

Bull, S., & Brna, P. (1999). Enhancing peer interaction in the Solar System. In *C-LEMMAS: Roles of Communicative Interaction in Learning to Model in Mathematics and Science*, Ajjacio Corsica.

Bull, S., & Shurville, S. (1999). Reader, writer and student models to support the writing process. *Proceedings of UM '99, 7th International Conference on User Modeling*, Banff, Canada. (pp. 295-297).

Bull, S., & Smith, M. (1997). A pair of student models to encourage collaboration. *User Modeling: Proceedings of the Sixth International Conference,* UM97 Vienna. New York, NY: Springer Wien.

Clancey, W. (1986). Qualitative student models. *Annual Review of Computer Science*, *1*, 381–450. doi:10.1146/annurev.cs.01.060186.002121

Collins, J. A., Greer, J. E., & Huang, S. X. (1996). Adaptive assessment using granularity hierarchies and bayenesian nets. In C. Frasson, G. Gauthier, & A. Lesgold (Eds.), *Proceedings of Intelligent Tutoring Systems (ITS'96),* Montreal Canada, (pp. 569-577).

Conati, C., Gertner, A., Van Lehn, K., & Druzdzel, M. (1997). Online student modeling for coached problem solving using Bayesian networks. In A. Jameson, C. Paris & C. Tasso (Eds.), *User Modeling: Proceedings of the Sixth International Conference,* UM97, (pp. 231-242). Vienna. New York, NY: Springer Wien.

Conati, C., & Van Lehn, K. (1999). A student model to assess self-explanation while learning from examples. *Proceedings of UM '99, 7th International Conference on User Modeling*, Banff, Canada. (pp. 303-305).

Conati, C., & Van Lehn, K. (1999). A student model to assess self-explanation while learning from examples. *Proceedings of UM '99, 7th International Conference on User Modeling*, Banff, Canada, (pp. 303-305).

Giangrandi, P., & Tasso, C. (1996). Modeling the temporal evolution of student's knowledge. In P. Brna, A. Paiva & J. Self (Eds.), *Proceedings of the European Conference on Artificial Intelligence in Education*. (pp. 184-190).

Giangrandi, P., & Tasso, C. (1997). Managing temporal knowledge in student modeling. *User Modeling: Proceedings of the Sixth International Conference*, UM97 Vienna. (pp. 415-426). New York, NY: Springer Wien.

Greer, J. (1996). *Student modelling*. User Modelling Conference, Kona, Hawaii USA.

Greer, J. E., & McCalla, G. I. (1994). *Student modelling: The key to individualized knowledge-based instruction. NATO ASI, Series F*. Springer-Verlag.

Holt, P., Dubs, S., Jones, M., & Greer, J. (1994). The state of student modelling. In Greer, J. E., & McCalla, G. I. (Eds.), *Student modelling: The key to individualized knowledge-based instruction* (pp. 3–35). NATO ASI, Series F., Springer-Verlag.

Kass, R. (1989). Student modelling in intelligent tutoring systems-implications for user modelling. In Kosba, A., & Wahlster, W. (Eds.), *User models in dialog systems* (pp. 386–410). Berlin, Germany: Springer-Verlag.

Matsuda, N., & Okamoto, T. (1992). Student model and its recognition by hypothesis-based reasoning in ITS. *Journal of Electronics and Communications in Japan*, *75*(8), 85–95. doi:10.1002/ecjc.4430750807

McCalla, G. (1992a). The central importance of student modelling to intelligent tutoring. In Costa, E. (Ed.), *New directions for intelligent tutoring systems*. Berlin, Germany: Springer Verlag.

McCalla, G. (1992b). The centrality of student modelling to intelligent tutoring. In Costa, E. (Ed.), *NATO ASI, Series F* (*Vol. 91*, pp. 107–131). Berlin, Germany: Springer-Verlag.

McCalla, G. I., & Greer, J. E. (1994). Granularity-based reasoning and belief revision in student models. In *Student models: The key to individualized educational systems* (pp. 39–62). New York, NY: Springer Verlag.

Nwana, D. (1991). User modeling and user adapted interaction in an intelligent tutoring system. *User Modeling and User-Adapted Interaction*, *1*, 1–32. doi:10.1007/BF00158950

Ragnemalm, E. L. (1999). *Student modelling based on collaborative dialogue with a learning companion.* Dissertation No 563, Linkoping University, S-581 83, Linkoping, Sweden.

Self, J. (1994). Formal approaches to student modeling. In McCalla, G., & Greer, J. (Eds.), *Student models: The key to individualized educational systems*. New York, NY: Springer Verlag.

Sison, R., & Shimura, M. (1998). Student modeling and machine learning international. *Journal of Artificial Intelligence in Education, 9*, 128–158.

VanLehn, K. (1988). Student modeling. In Polson, M. C., & Richardson, J. J. (Eds.), *Foundations of intelligent tutoring systems* (pp. 55–78).

VanLehn, K., & Martin, J. (1997). Evaluation of an assessment system based on Bayesian student modelling, 8. (pp. 179-221).

Compilation of References

ACG/CEG. (2008). *The SOLERES R&D Project: A spatio-temporal environmental management Information System based on neural-networks, agents and software components*. (Technical report, Applied Computing Group (ACG) and Computers and Environmental Group (CEG), University of Almeria, Spain). Retrieved from http://www.ual.es/acg/soleres

Adler, P., & Kwon, S. (2002). Social capital: Prospects for a new concept. *Academy of Management Review*, *27*, 17–40. doi:10.2307/4134367

Ahn, H., & Kim, K. J. (2009). Global optimization of case-based reasoning for breast cytology diagnosis. *Expert Systems with Applications*, *36*(1), 724–734. doi:10.1016/j.eswa.2007.10.023

Aho, A., Chang, S., McKeown, K., Radev, D., Smith, J., & Zaman, K. (1997). Columbia Digital news project: An environment for briefing and search over multimedia. *International Journal on Digital Libraries*, *1*(4), 377–385. doi:10.1007/s007990050030

Alavi, M., & Leidner, D. E. (2001). Knowledge management and knowledge management systems: Conceptual foundations and research issues. *Management Information Systems Quarterly*, *25*(1), 107–136. doi:10.2307/3250961

Alayón, S., Robertson, R., Warfield, S. K., & Ruiz-Alzola, J. (2007). A fuzzy system for helping medical diagnosis of malformations of cortical development. *Journal of Biomedical Informatics*, *40*(3), 221–235. doi:10.1016/j.jbi.2006.11.002

Albert, M. (2000). *Quantum mechanics*. New York, NY: Dover Publications.

Albert, R., & Barabasi, A.-L. (2002). Statistical mechanics of complex networks. *Reviews of Modern Physics*, *74*, 47–94. doi:10.1103/RevModPhys.74.47

Aleksovski, Z., Klein, M., Kate, W., & Van Harmelen, F. (2006). Matching unstructured vocabularies using a background ontology. In Staab, S., & Svátek, V. (Eds.), *EKAW 2006. (LNCS/LNAI 4248)* (pp. 182–197).

Amaral, L. A. N., & Ottino, J. M. (2004). Complex networks-augmenting the framework for the study of complex systems. *The European Physical Journal*, *38*, 147–162.

Ambler, S. (2003). *Agile database techniques, effective strategies for the agile software developer*. Wiley.

Ambler, S. W. (2002). *Agile modelling, effective practices for extreme programming and the unified process*. John Wiley & Sons Inc.

Anderson, J., Corbett, A., Koedinger, K., & Pelletier, R. (1995). Cognitive tutors: Lessons learned. [NJ: LEA.]. *Journal of the Learning Sciences*, *4*(2), 167–207. doi:10.1207/s15327809jls0402_2

Android. (2010). *Specifications*. Retrieved from http://developer.android.com/

Andronico, A., Carbonaro, A., Colazzo, L., Molinari, A., & Ronchetti, M. (2003). *Designing models and services for learning management systems in mobile settings*. In Mobile and Ubiquitous Information Access: Mobile HCI 2003 International Workshop, (LNCS 2954), (pp. 90-106). ISBN 3-540-21003-2

Angehrn, A. A., Maxwell, K., Luccini, A. M., & Rajola, F. (2008). Designing collaborative learning and innovation systems for education professionals. In Miltiadis, D., Lytras, J. M., Carroll, E. D., & Tennyson, R. D. (Eds.), *Emerging technologies and Information Systems for the knowledge society. (LNCS 5288)* (pp. 167–176). Heidelberg, Germany: Springer. doi:10.1007/978-3-540-87781-3_19

Anti-phishing Working Group. (2009). http://www.antiphishing.org

Asensio, J. A., Iribarne, L., Padilla, N., & Ayala, R. (2008). Implementing trading agents for adaptable and evolutive COTS components architectures. In *Proceedings of the International Conference on e-Business* (pp. 259-262). Porto, Portugal.

Assenova, P., & Johannesson, P. (1996). *Improving quality in conceptual modelling by the use of schema transformations*. 15th International Conference on Conceptual Modeling.

Astels, D. (2002). *Refactoring with UML*. Conference on eXtreme Programming and Agile Processes in Software Engineering XP2002, (pp. 67-70).

Auberger, M. (1988). *Student modeling in educational multimedia titles using agents*. Department of Information Business, Vienna University. Retrieved from http://lux2.wu-wien.ac.at/~geyers/archive/auberger/proposal/proposal.html

Aussenac-Gilles, N., Biébow, B., & Szulman, S. (2000). Revisiting ontology design: A method based on corpus analysis. *Proceedings of the 12th International Conference on Knowledge Engineering and Knowledge Management, methods, models and tools* (LNAI 1937), (pp. 172-188). Heldelberg, Germany: Springer-Verlag.

Avesani, P., Massa, P., & Tiella, R. (2005). A trust-enhanced recommender system application: Moleskiing. *Proceedings of the 2005 ACM Symposium on Applied Computing* (SAC), (pp.1589-1593).

Avilés, W., Ortega, O., Kuan, G., Coloma, J., & Harris, E. (2008). Quantitative assessment of the benefits of specific information technologies applied to clinical studies in developing countries. *The American Journal of Tropical Medicine and Hygiene*, *78*(2), 311–315.

Baffes, P., & Mooney, R. (1996). Refinement-based student modeling and automated bug library construction. *Journal of Artificial Intelligence in Education*, *7*(1), 75–116.

Balabanovic, M., & Shoham, Y. (1997). Content-based, collaborative recommendation. *Communications of the ACM*, *40*(3), 66–72. doi:10.1145/245108.245124

Barabasi, A.-L., & Réka, A. (1999). Emergence of scaling in random networks. *Science*, *286*, 509–512. doi:10.1126/science.286.5439.509

Barla, M., & Bieliková, M. (2007). Estimation of user characteristics using rule-based analysis of user logs. *Proceedings of Data Mining for User Modeling Workshop held at the International Conference on User Modeling*, (pp. 5-14).

Barr, A., Beard, M., & Atkinson, R. C. (1982). The computer as a tutorial laboratory: The Stanford BIP Project. *International Journal of Man-Machine Studies*, *8*, 567–596. doi:10.1016/S0020-7373(76)80021-1

Barzilay, R., McKeown, K., & Elhadad, M. (1999). Information fusion in the context of multi-document summarization. In *Proceedings of ACL'99*.

Batini, C., & Ceri, S. (1992). *Conceptual database design, an entity-relationship approach*. Benjamin/Cummings.

Beck, K. (1999). *Extreme programming explained*. Addison-Wesley Pub Co.

Beck, J., Stern, M., & Woolf, B. P. (1997). Using the student model to control problem difficulty. *User Modeling: Proceedings of the Sixth International Conference*, UM97 Vienna. (pp. 277-288). New York, NY: Springer Wien.

Beckett, D. (Ed.). (2004). *RDF/XML syntax specification*. Retrieved May 12, 2010, from http://www.w3.org/TR/rdf-syntax-grammar/

Beged-Dov, G., Bricley, D., Dornfest, R., Davis, I., Dodds, L., & Eisenzopf, J. … E. van der Vlist (Eds.). (2001). *RDF site summary (RSS) 1.0*. Retrieved May 15, 2010, from http://web.resource.org/rss/1.0/spec

Belkin, N. J., & Croft, W. B. (1992). Information filtering and information retrieval: Two sides of the same coin? *Communications of the ACM*, *35*(12), 29–38. doi:10.1145/138859.138861

Benbya, H. (2008). *Knowledge management systems implementation: Lessons from the Silicon Valley*. Oxford, UK: Chandos Publishing.

Berendt, B. (2007). Context, (e)learning, and knowledge discovery for Web user modelling: Common research themes and challenges. *Proceedings of Data Mining for User Modeling Workshop held at the International Conference on User Modeling,* (pp. 15-27).

Bergstein, P. L. (1991). Object-preserving class transformations. *Conference proceedings on Object-oriented programming systems, languages, and applications*. Phoenix, Arizona, United States, ACM Press, (pp. 299-313).

Berliant, M., & Fujita, M. (2007). *Knowledge creation as a square dance on the Hilbert cube*. (MPRA Paper No. 4680). Retrieved from http://mpra.ub.uni-muenchen.de/4680/

Berners-Lee, T., Hendler, J., & Lassila, O. (2001, May). The Semantic Web: A new form of Web content that is meaningful to computers will unleash a revolution of new possibilities. *The Scientific American*. Retrieved May 10, 2010, from http://www.sciam.com/article.cfm?id=the-semantic-web

Bernes-Lee, T. (2000). *What the Semantic Web can represent*. Retrieved from http://www.w3.org/DesignIssues/RDFnot.html

Beudoin, C. E. (2008). Explaining the relationship between Internet use and interpersonal trust: Taking into account motivation and information overload. *Journal of Computer-Mediated Communication, 13*, 550–568. doi:10.1111/j.1083-6101.2008.00410.x

Bighini, C., & Carbonaro, A. (2004). InLinx: Intelligent agents for personalized classification, sharing and recommendation. *International Journal of Computational Intelligence, 2*(1).

Bighini, C., Carbonaro, A., & Casadei, G. (2003). Inlinx for document classification, sharing and recommendation. In V. Devedzic, J. M. Spector, D. G. Sampson, & Kinshuk (Eds.), *Proceedings of the 3rd Int'l. Conf. on Advanced Learning Technologies*, (pp. 91–95). Los Alamitos, CA: IEEE Computer Society.

Billsus, D., & Pazzani, M. J. (1999). A hybrid user model for news story classification. In *Proceedings of the 7th International Conference on User Modeling* (UM'1999), (pp. 99- 108), Banff, Canada: Springer-Verlag, New York, Inc.

Bluetooth Core Specification v3.0 + HS (2009). *Specifications*. Retrieved from http://www.bluetooth.com/Bluetooth/Technology/Building/Specifications/

Bois, B. D. (2004). Opportunities and challenges in deriving metric impacts from refactoring postconditions. In S. Demeyer, S. Ducasse & K. Mens (Eds.), *Proceedings WOOR'04, ECOOP'04 Workshop on Object-Oriented Re-engineering*. Antwerpen, Belgium.

Bois, B. D., Demeyer, S., et al. (2004). Refactoring-improving coupling and cohesion of existing code. In *Proceedings WCRE'04 Working Conference on Reverse Engineering*. Victoria, Canada, IEEE Press, (pp. 144-151).

Borgman, C. L. (2007). *Scholarship in the digital age: Information, infrastructure, and the Internet*. Cambridge, MA: MIT Press.

Bottoni, P., Parisi-Presicce, F., et al. (2002). Coordinated distributed diagram transformation for software evolution. In R. Heckel, T. Mens & M. Wermelinger (Eds.), Proceedings of the Workshop on 'Software Evolution Through Transformations'(SET'02), (p. 72).

Bouzeghoub, A., Carpentier, C., Defude, B., & Duitama, J. F. (2003). A model of reusable educational components for the generation of adaptive courses. In *Proceedings of the First International Workshop on Semantic Web for Web-Based Learning at CAISE'03*, Klagenfurt, Austria. Retrieved from http://www-inf.int-evry.fr/~defude/PUBLI/SWWL03.pdf

Bowne. (2000). *Open sesame 2000*. Retrieved from http://www.opensesame.com

Brehm, J., & Rahn, W. (1997). Individual-level evidence for the causes and consequences of social capital. *American Journal of Political Science, 41*(3), 999–1024. doi:10.2307/2111684

Brewer, C. A. (2006). Basic mapping principles for visualizing cancer data uUsing Geographic Information Systems (GIS). *American Journal of Preventive Medicine, 30*(2), S25–S36. doi:10.1016/j.amepre.2005.09.007

Brunn, M., Chali, Y., & Pinchak, C. (2001). *Text summarization using lexical chains.*

Brusilovsky, P. (1996). Methods and techniques of adaptive hypermedia. *User Modeling and User-Adapted Interaction, 6*(2-3), 87–129. doi:10.1007/BF00143964

Budanitsky, A., & Hirst, G. (2001). *Semantic distance in Wordnet: An experimental, application-oriented evaluation of five measures.* In Workshop on WordNet and Other Lexical Resources. Second meeting of the North American Chapter of the Association for Computational Linguistics, Pittsburgh.

Bull, S. (1997). See yourself write: A simple student model to make students think. *User Modeling: Proceedings of the Sixth International Conference*, UM97 Vienna. (pp. 315-326). New York, NY: Springer Wien.

Bull, S., & Brna, P. (1999). Enhancing peer interaction in the Solar System. In *C-LEMMAS: Roles of Communicative Interaction in Learning to Model in Mathematics and Science*, Ajjacio Corsica.

Bull, S., & Shurville, S. (1999). Reader, writer and student models to support the writing process. *Proceedings of UM '99, 7th International Conference on User Modeling*, Banff, Canada. (pp. 295-297).

Bull, S., & Smith, M. (1997). A pair of student models to encourage collaboration. *User Modeling: Proceedings of the Sixth International Conference,* UM97 Vienna. New York, NY: Springer Wien.

Burke, R. (2007). Hybrid Web recommender systems. In Brusilovsky, P., Kobsa, A., & Nejdl, W. (Eds.), *The adaptive Web: Methods and srategies of Web personalization* (pp. 377–408). Heidelberg, Germany: Springer Verlag.

Cao, Y., & Li, Y. (2007). An intelligent fuzzy-based recommendation system for consumer electronic products. [Amsterdam, The Netherlands: Elsevier.]. *Expert Systems with Applications, 33*(1), 230–240. doi:10.1016/j.eswa.2006.04.012

Carbonaro, A. (2006). Defining personalized learning views of relevant learning objects in a collaborative bookmark management system. In Ma, Z. (Ed.), *Web-based intelligent e-learning systems: Technologies and applications* (pp. 139–155). Hershey, PA: Information Science Publishing.

Carbonaro, A. (2010). *Towards an automatic forum summarization to support tutoring.* In M. D. Lytras, P. Ordonez de Pablos, D. Avison, J. Sipior, Q. Jin, W. Leal...D. G. Horner (Eds.), *Technology enhanced learning: Quality of teaching and educational reform, communications in computer and Information Science*, vol. 73. (pp. 141-147). ISBN: 978-3-642-13165-3

Carbonaro, A., & Ferrini, R. (2005). *Considering semantic abilities to improve a Web-based distance learning system.* ACM International Workshop on Combining Intelligent and Adaptive Hypermedia Methods/Techniques in Web-based Education Systems.

Carcagni, A., Corallo, A., Zilli, A., Ingraffia, N., & Sorace, S. (2008). A workflow management system for ontology engineering. In Zilli, A., Damiani, E., Ceravolo, P., Corallo, A., & Elia, G. (Eds.), *Semantic knowledge management: An ontology-based framework.* Hershey, PA: IGI Global.

Carrillo-Ramos, A., Gensel, J., Villanova-Oliver, M., & Martin, H. (2006). *Adapted information retrieval in Web Information Systems using PUMAS.* (LNCS 3529), (p. 243).

Casais, E. (1989). Reorganizaing an object system. In Tsichritzis, D. (Ed.), *Object oriented development* (pp. 161–189). Geneve.

Castro, F., Vellido, A., Nebot, A., & Mugica, F. (2005). Applying data mining techniques to e-learning problems. [SCI]. *Studies in Computational Intelligence, 62*, 183–221. doi:10.1007/978-3-540-71974-8_8

Ceri, S., Daniel, F., Matera, M., & Facca, F. M. (2007). Model-driven development of context-aware Web applications. *ACM Transactions on Internet Technology, 7*(1). doi:10.1145/1189740.1189742

Chaffee, J., & Gauch, S. (2000). Personal ontologies for Web navigation. In *Proceedings of the 9th International Conference on Information and Knowledge Management* (CIKM '2000), (pp. 227-234). McLean, VA/ New York, NY: ACM.

Chang, Y., Chen, C. S., & Zhou, H. (2009). Smart phone for mobile commerce. *Computer Standards & Interfaces, 31*(4), 740–747. doi:10.1016/j.csi.2008.09.016

Chen, L., & Sycara, K. (1998). A personal agent for browsing and searching. In *Proceedings of the 2nd International Conference on Autonomous Agents* (pp. 132-139). Minneapolis/St. Paul.

Chien, B.-C., Hu, C.-H., & Ju, M.-Y. (2010). Ontology-based information retrieval using fuzzy concept documentation. *Cybernetics and Systems*, *41*(1), 4–16. doi:10.1080/01969720903408565

Chien, B.-C., Hu, C.-H., & Ju, M.-Y. (2007). Intelligent information retrieval applying automatic constructed fuzzy ontology. In *proceedings of 6th International Conference on Machine Learning and Cybernetics*, 1-7 (pp. 2239-2244). Hong Kong.

Chung, L., & Nixon, B. A. (2000). *Non-functional requirements in software engineering*. Kluwer Academic Publishers.

Cilibrasi, R., & Vitányi, P. (2007). The Google similarity distance. *IEEE Transactions on Knowledge and Data Engineering*, *19*(3), 370–383. doi:10.1109/TKDE.2007.48

Cisternino, V., Campi, E., Corallo, A., Taifi, N., & Zilli, A. (2008). Ontology-based knowledge management systems for the new product development acceleration: Case of a community of designers of automotives. *IEEE Proceedings of the 1st KARE Workshop in the 4th SITIS*, Nov. 30- Dec. 3, 2008, Bali, Indonesia.

Clancey, W. (1986). Qualitative student models. *Annual Review of Computer Science*, *1*, 381–450. doi:10.1146/annurev.cs.01.060186.002121

Clarke, F., & Ekeland, I. (1982). Nonlinear oscillations and boundary-value problems for Hamiltonian systems. *Archive for Rational Mechanics and Analysis*, *78*, 315–333. doi:10.1007/BF00249584

Cleverdon, C. W., Mills, J., & Keen, E. M. (1966). *Factors determining the performance of indexing systems*, vol. 2, test results. Cranfield: ASLIB Cranfield Project.

Clippinger, J. H. III. (1999). *The biology of business. Decoding the natural laws of enterprise*. San Francisco, CA: Jossey-Bass Publishers.

Collins, J. A., Greer, J. E., & Huang, S. X. (1996). Adaptive assessment using granularity hierarchies and bayenesian nets. In C. Frasson, G. Gauthier, & A. Lesgold (Eds.), *Proceedings of Intelligent Tutoring Systems (ITS'96)*, Montreal Canada, (pp. 569-577).

Colquhoun, G. J., Baines, R. W., & Crosseley, R. (1993). A state of the art review of IDEF0. *International Journal of Computer Integrated Manufacturing*, *6*(4), 252–264. doi:10.1080/09511929308944576

Conati, C., & Van Lehn, K. (1999). A student model to assess self-explanation while learning from examples. *Proceedings of UM '99, 7th International Conference on User Modeling*, Banff, Canada. (pp. 303-305).

Conati, C., Gertner, A., Van Lehn, K., & Druzdzel, M. (1997). Online student modeling for coached problem solving using Bayesian networks. In A. Jameson, C. Paris & C. Tasso (Eds.), *User Modeling: Proceedings of the Sixth International Conference*, UM97, (pp. 231-242). Vienna. New York, NY: Springer Wien.

Conesa, J. (2004). *Ontology-driven Information Systems, pruning and refactoring of ontologies*. Doctoral Symposium of the Seventh International Conference on the Unified Modeling language. Retrieved from http.//www-ctp.di.fct.unl.pt/UML2004/DocSym/JordiConesaUML2004DocSym.pdf

Conesa, J. (2008). *Pruning and refactoring ontologies in the development of conceptual schemas of Information Systems*. Unpublished doctoral thesis. Technical University of Catalonia.

Cooper, R. G. (1994). Perspectives: Third-generation new product processes. *Journal of Product Innovation Management*, *11*, 3–14. doi:10.1016/0737-6782(94)90115-5

Corallo, A., Laubacher, R., Margherita, A., & Turrisi, G. (2009). Enhancing product development through knowledge-based engineering (KBE): A case study in the aerospace industry. *Journal of Manufacturing Technology Management*, *20*(8), 1070–1083. doi:10.1108/17410380910997218

Corallo, A., Elia, G., & Zilli, A. (2005). Enhancing communities of practice: An ontological approach. *Proceedings of the 11th International Conference on Industrial Engineering and Engineering Management*, 23-25 April, Northeastern University, Shenyang, China.

Corallo, A., Lazoi, M., Taifi, N., & Passiante, G. (2010). Integrated systems for product design: The move toward outsourcing. *Proceedings of the IADIS International Conference on Information Systems*, March, 18-21, Porto, Portugal.

Corallo, A., Taifi, N., & Passiante, G. (2008). Strategic and managerial ties of the new product development. *CCIS Proceedings of the First World Summit of the Knowledge Society, 19*, (pp. 398-405). Sept. 24-26, Athens, Greece.

Corbett, A. T., & Anderson, J. R. (1994). Knowledge tracing: Modeling the acquisition of procedural knowledge. *User Modeling and User-Adapted Interaction, 4*(4), 253–278. doi:10.1007/BF01099821

Correa, A., & Werner, C. (2004). Applying refactoring techniques to UML/OCL models. *Proceedings of International Conference of UML*. Lisbon, Portugal, (pp. 173-187).

Corritore, C., Kracher, B., & Wiedenbeck, S. (2003). Online trust: Concepts, evolving themes, a model. *International Journal of Human-Computer Studies, 58*(6), 737–758. doi:10.1016/S1071-5819(03)00041-7

Cretton, F., & Calvé, A. L. (2007). *Working paper: Generic ontology based user modeling-GenOUM*. (Technical report, HES-SO Valais).

Critchlow, M., Dodd, K., et al. (2003). Refactoring product line architectures. *Proceedings of the First International Workshop on REFactoring, Achievements, Challenges, Effects* (REFACE). Victoria, Canada.

Cunningham, H., Maynard, D., Bontcheva, K., & Tablan, V. (2002). GATE: A framework and graphical development environment for robust NLP tools and applications. In *Proceedings 40th Anniversary Meeting of the Association for Computational Linguistics* (ACL 2002). Budapest.

Damiani, E., Ceravolo, P., Corallo, A., Elia, G., & Zilli, A. (2009). KIWI: A framework for enabling semantic knowledge management. In Zilli, A., Damiani, E., Ceravolo, P., Corallo, A., & Elia, G. (Eds.), *Semantic knowledge management: An ontology-based framework*. Hershey, PA: IGI Global.

Dartigues, C., Ghodous, P., Gruninger, M., Pallez, D., & Sriram, R. (2007). CAD/CAPP integration using feature ontology. *Concurrent Engineering-Research and Applications, 15*(2), 237–249. doi:10.1177/1063293X07079312

Davenport, T., & Grover, V. (2001). Knowledge management. *Journal of Management Information Systems, 18*(1), 3–4.

De Bra, P., Aerts, A., Berden, B., de Lange, B., Rousseau, B., Santic, T., et al. (2003). AHA! The adaptive hypermedia architecture. In *Proceedings of the Fourteenth ACM Conference on Hypertext and Hypermedia* (pp. 81-84). Nottingham, UK: ACM Press.

De Leenheer, P., & Mens, T. (2008). Ontology evolution: State of the art & future directions. In Hepp, M., De Leenheer, P., de Moor, A., & Sure, Y. (Eds.), *Ontology management for the Semantic Web, Semantic Web services, and business applications, from Semantic Web and beyond: Computing for human experience*. Springer.

De Leenheer, P. (2008). Keynote: Towards an ontological foundation for evolving agent communities. *Proceedings of 32nd Annual IEEE International Computer Software and Applications Conference*, (pp. 523-528).

Denaux, R., Aroyo, L., & Dimitrova, V. (2005a). An approach for ontology-based elicitation of user models to enable personalization on the Semantic Web. In *Proceedings of the 14th international conference on World Wide Web* (WWW '05), (pp. 1170-1171), Chiba, Japan. New York, NY: ACM.

Desimone, L. (1999). Linking parent involvement with student achievement: Do race and income matter? *The Journal of Educational Research, 93*(1), 11–30. doi:10.1080/00220679909597625

Deursen, A. v., Marin, M., et al. (2003). Aspect mining and refactoring. *Proceedings of the First International Workshop on REFactoring, Achievements, Challenges, Effects* (REFACE03). Victoria, Canada.

Devedzic, V. (2001). Knowledge modeling-state of the art. *Integrated Computer-Aided Engineering, 8*(3), 257–281.

Dey, A. K., & Abowd, G. D. (2000). *Towards a better understanding of context and context-awareness*. In CHI 2000 Workshop on the What, Who, Where, When, and How of Context-Awareness.

Du, H. S., & Wagner, C. (2006). Weblog success: Exploring the role of technology. *International Journal of Human-Computer Studies*, *64*(9), 789–798. doi:10.1016/j.ijhcs.2006.04.002

Ducasse, S. e., Rieger, M., et al. (1999). A language independent approach for detecting duplicated code. In H. Yang & L. White (Eds.), *Proceedings ICSM'99 International Conference on Software Maintenance*. IEEE, (pp. 109-118).

Ducasse, S. e., Rieger, M., et al. (1999). *Tool support for refactoring duplicated OO code*. ECOOP'99, Springer. 1743, (pp. 177-178).

Dutton, W. H., & Helsper, E. (2007). *Oxford Internet survey 2007 report: The Internet in Britain*. Oxford, UK: Oxford Internet Institute.

Dutton, W. H., Helsper, E. J., & Gerber, M. M. (2009). *Oxford Internet survey 2009 report: The Internet in Britain*. Oxford Internet Institute, University of Oxford.

Dutton, W. H., & Shepherd, A. (2006). Trust in the Internet as an experience technology. *Information Communication and Society*, *9*, 433–451. doi:10.1080/13691180600858606

Eetvelde, N. V., & Janssens, D. (2004). *A hierarchical program representation for refactoring*. (ENTS 82).

Eick, C. F. (1991). *A methodology for the design and transformation of conceptual schemas*. Very Large Data Bases Conference, Barcelona.

Eisert, J., & Wilkens, M. (2000). Quantum games. *Journal of Modern Optics*, *47*, 25–43.

E-LIS. (2009). *Home page*. Retrieved May 10, 2010, from http://eprints.rclis.org/

Epstein, J. M., & Axtell, R. (1996). *Growing artificial societies*. Washington, D.C.: Brookings Institution Press.

Ernst, M. D., & Cockrell, J. (2001). Dynamically discovering likely program invariants to support program evolution. *IEEE Transactions on Software Engineering*, *27*(2), 1–25. doi:10.1109/32.908957

European Commission. (2005). *The business case for diversity. Good practices in the workplace, Belgium*. Retrieved from http://ec.europa.eu/social/main.jsp?catId=370&langId=en&featuresId=25

Euzenat, J., & Shvaiko, P. (2004). A survey of schema-based matching approaches. *Journal on Data Semantics*, *4*, 146–171.

Euzenat, J., & Shvaiko, P. (2007). *Ontology matching*. Berlin/ Heidelberg, Germany: Springer-Verlag.

Eysenbach, G. (2008). Medicine 2.0: Social networking, collaboration, participation, apomediation, and openness. *Journal of Medical Internet Research*, *10*(3), e22. doi:10.2196/jmir.1030

Falkman, G., Gustafsson, M., Jontell, M., & Torgersson, O. (2008). SOMWeb: A Semantic Web-based system for supporting collaboration of distributed medical communities of practice. *Journal of Medical Internet Research*, *10*(3), e25. doi:10.2196/jmir.1059

Farell, R. (2002). Summarizing electronic discourse. *International Journal of Intelligent Systems in Accounting Finance & Management*, *11*, 23–38. doi:10.1002/isaf.211

Faria, S., Fernandes, T. R., & Perdigoto, F. S. (2008). *Mobile Web server for elderly people monitoring*. ISCE 2008, IEEE International Symposium on Consumer Electronics.

Fellbaum, C. (Ed.). (1998). *WordNet: An electronic lexical database*. Cambridge, MA: MIT Press.

Fensel, D., van Harmelen, F., Horrocks, I., McGuinness, D. L., & Patel-Schneider, P. F. (2001). OIL: An ontology infrastructure for the Semantic Web. *IEEE Intelligent Systems*, *16*(2), 38–45. doi:10.1109/5254.920598

Fink, J., & Kobsa, A. (2000). A review and analysis of commercial user modeling servers for personalization on the World Wide Web. *International Journal on User Modeling and User-Adapted Interaction*, *10*(2-3), 209–249. doi:10.1023/A:1026597308943

Fink, J., & Kobsa, A. (2002). User modeling in personalized city tours. *Artificial Intelligence Review*, *18*(1), 33–74. doi:10.1023/A:1016383418977

Flouris, G. (2007). Ontology change: Classification and survey. [United Kingdom: Cambridge University Press.]. *The Knowledge Engineering Review*, *23*(2), 117–152.

Fowler, M. (1999). *Refactoring, improving the design of existing code*. Addison-Wesley.

Fowler, M., & Sadalage, P. (2003). *Evolutionary database design*. Retrieved from http.//www.martinfowler.com/articles/evodb.html

Frias-Martinez, E., Chen, S., & Liu, X. (2006). Survey of data mining approaches to user modeling for adaptive hypermedia. *IEEE Transactions on Systems, Man and Cybernetics. Part C, Applications and Reviews*, *36*(6), 734–749. doi:10.1109/TSMCC.2006.879391

Friedman, M., Resnick, P., & Sami, R. (2007). *Algorithmic game theory*. Cambridge, UK: Cambridge University Press.

Fuentes-Lorenzo, D., Morato, J., & Gómez-Berbís, J. M. (2009). Knowledge management in biomedical libraries: A Semantic Web approach. *Information Systems Frontiers*, *11*(4), 471–480. doi:10.1007/s10796-009-9159-y

Füller, J., Bartl, M., Ernst, H., & Mühlbacher, H. (2006). Community-based innovation: How to integrate members of virtual communities into new product development. *Electronic Commerce Research*, *6*, 57–73. doi:10.1007/s10660-006-5988-7

Galagher, S. (2009). *Social networks a magnet for malware*. IT management e-book. WebMediaBrands.

Gamma, E., & Helm, R. (1995). *Design patterns, elements of reusable object-oriented software*. Addison-Wesley.

García-Crespo, A., Chamizo, J., Colomo-Palacios, R., Mendoza-Cembranos, M. D., & Gómez-Berbís, J. M. (2010). (in press). S-SoDiA: A semantic enabled social diagnosis advisor. *International Journal of Society Systems Science*.

García-Crespo, A., Rodriguez, A., Mencke, M., Gómez-Berbís, J. M., & Colomo Palacios, R. (2010). ODDIN: Ontology-driven differential diagnosis based on logical inference and probabilistic refinements. *Expert Systems with Applications*, *37*(3), 2621–2628. doi:10.1016/j.eswa.2009.08.016

García-Sánchez, F., Fernández-Breis, J. T., Valencia-García, R., Gómez, J. M., & Martínez-Béjar, R. (2008). Combining Semantic Web technologies with multi-agent systems for integrated access to biological resources. *Journal of Biomedical Informatics*, *41*(5), 848–859. doi:10.1016/j.jbi.2008.05.007

Germain, R. (1996). The role of context and structure in radical and incremental logistics innovation adoption. *Journal of Business Research*, *35*, 117–127. doi:10.1016/0148-2963(95)00053-4

Giangrandi, P., & Tasso, C. (1996). Modeling the temporal evolution of student's knowledge. In P. Brna, A. Paiva & J. Self (Eds.), *Proceedings of the European Conference on Artificial Intelligence in Education*. (pp. 184-190).

Giangrandi, P., & Tasso, C. (1997). Managing temporal knowledge in student modeling. *User Modeling: Proceedings of the Sixth International Conference*, UM97 Vienna. (pp. 415-426). New York, NY: Springer Wien.

Gibson, R., & Schuyler, E. (2006). *Google maps hacks: Tips & tools for geographic searching and remixing* (hacks). Sebastopol, CA: O'Reilly Media Inc.

Girvan, M., & Newman, M. E. J. (2002). Community structure in social and biological

Giustini, D. (2006). How Web 2.0 is changing medicine. *British Medical Journal*, *333*, 1283–1284. doi:10.1136/bmj.39062.555405.80

Giustini, D. (2007). Web 3.0 and medicine. Make way for the Semantic Web. *British Medical Journal*, *335*, 1273–1274. doi:10.1136/bmj.39428.494236.BE

Global Trends 2025: A transformed world. (2008). *Appendix F: The Internet of things* (background). SRI Consulting Business Intelligence.

Gloor, P., Grippa, F., Kidane, Y. H., Marmier, P., & Von Arb, C. (2008). Location matters-measuring the efficiency of business social networking. *International Journal of Foresight and Innovation Policy*, *4*(3/4), 230–245. doi:10.1504/IJFIP.2008.017578

Golbeck, J. (2007). The dynamics of Web-based social networks: Membership, relationships, and change. *First Monday*, *12*(11).

Golbeck, J. (2008). *Trust on the World Wide Web: A survey*. Hanover, MA: NowPublishers.

Golbeck, J., & Hendler, J. (2004). Reputation network analysis for email filtering.

Goodwin, J., & Emirbayer, M. (1999). Network analysis, culture, and the problem of agency. *American Journal of Sociology*.

Gorp, P. V., Stenten, H., et al. (2003). Towards automating source-consistent UML refactorings. UML, 2003, Springer. 2863, (pp. 144-158).

Grabler, F., Agrawala, M., Sumner, R. W., & Pauly, M. (2008). Automatic generation of tourist maps. *ACM Transactions on Graphics*, *27*(3). doi:10.1145/1360612.1360699

Grant, R. (1996). Towards a knowledge based theory of the firm. *Strategic Management Journal*, *17*(1), 109–122.

Grant, S., & Cordy, J. R. (2003). An interactive interface for refactoring using source transformation. *Proceedings of the Fist International Workshop on Refactoring, Achievements, Challenges and Effects* (REFACE'03). Victoria, Canada, (pp. 30-33).

Gray, J. G. (2002). *Aspect-oriented domain-specific modeling, a generative approach using a meta-weaver framework. Department of Electrical Engineering and Computer Science*. Nashville: Vanderbilt University.

Greer, J. E., & McCalla, G. I. (1994). *Student modelling: The key to individualized knowledge-based instruction. NATO ASI, Series F*. Springer-Verlag.

Greer, J. (1996). *Student modelling*. User Modelling Conference, Kona, Hawaii USA.

Griswold, W. G., & Chen, M. I. (1998). Tool support for planning the restructuring of data abstractions in large systems. *IEEE Transactions on Software Engineering*, *24*(7), 534–558. doi:10.1109/32.708568

Griswold, W. G., & Notkin, D. (1993). Automated assistance for program restructuring. *ACM Transactions on Software Engineering and Methodology*, *2*(3), 228–269. doi:10.1145/152388.152389

Gruber, T. R. (1995). Toward principles for the design of ontologies used for knowledge sharing. *International Journal of Human-Computer Studies*, *43*(5-6), 907–928. doi:10.1006/ijhc.1995.1081

Gruber, T. R. (2008). Collective knowledge systems: Where the social Web meets the Semantic Web. *Web Semantics: Science. Services and Agents on the World Wide Web*, *6*(1), 4–13.

Guarino, N. (1997). Understanding building and using ontologies: A commentary to using explicit ontologies in KBS development by van Heijst, Shreiber, & Wielinga. *International Journal of Human-Computer Studies*, *46*, 293–310. doi:10.1006/ijhc.1996.0091

Guarino, N. (1998). Formal ontology and Information Systems. In Guarino, N. (Ed.), *Formal ontology in Information Systems* (pp. 3–17). Amsterdam, The Netherlands: IOS Press.

Gunawardena, C. N., & Zittle, F. J. (1997). Social presence as a predictor of satisfaction within a computer mediated conferencing environment. *American Journal of Distance Education*, *11*(3), 8–26. doi:10.1080/08923649709526970

Halpin, T. (1996). *Conceptual schema and relational database design*. Prentice-Hall, Inc.

Halpin, T. (2001). *Information modeling and relational: From conceptual analysis to logical design*. Morgan Kaufman.

Halpin, T. A., & Proper, H. A. (1995). *Database schema transformation & optimization*. International Conference on Conceptual Modeling, LNCS.

Hansen, M. (1999). The search-transfer problem: the role of weak ties in sharing knowledge across organization subunits. *Administrative Science Quarterly*, *44*(1), 82–11. doi:10.2307/2667032

Hearst, M. A. (1992). Automatic acquisition of hyponyms from large text corpora. *Proceedings of the 14th conference on Computational Linguistics*, New Jersey, USA, (pp. 539-545).

Heckmann, D., Schwarzkopf, E., Mori, J., Dengler, D., & Kröner, A. (2007). The user model and context ontology GUMO revisited for future Web 2.0 extensions. *Proceedings of Contexts and Ontologies: Representation and Reasoning*, (pp. 37-46).

Hendler, J. (2001). Agents and the Semantic Web. *IEEE Intelligent Systems*, (March-April): 30–37. doi:10.1109/5254.920597

Herlocker, J.-L., Konstan, J.-A., & Riedl, J. (2000). Explaining collaborative filtering recommendations. *Proceedings of the 2000 ACM conference on Computer supported cooperative work*, Philadelphia, Pennsylvania, United States, (pp. 241-250).

Herrera, F., Herrera-Viedma, E., & Verdegay, J. L. (1996). Direct approach processes in group decision making using linguistic OWA operators. *Fuzzy Sets and Systems, 79*(2), 175–190. doi:10.1016/0165-0114(95)00162-X

Herrera, F., & Martinez, L. (2000). A 2-tuple fuzzy linguistic representation model for computing with words. *IEEE Transactions on Fuzzy Systems, 8*(6), 746–752. doi:10.1109/91.890332

Herrera-Viedma, E., Peis, E., Morales-del-Castillo, J. M., & Anaya, K. (2007). Improvement of Web-based service Information Systems using fuzzy linguistic techniques and Semantic Web technologies. In Liu, J., Ruan, D., & Zhang, G. (Eds.), *E-service intelligence: Methodologies, technologies and applications* (pp. 647–666). Heidelberg, Germany: Springer-Verlag.

Hoffman, D. L., & Novak, T. P. (2009). Flow online: Lessons learned and future prospects. *Journal of Interactive Marketing, 23*(1), 23–34. doi:10.1016/j.intmar.2008.10.003

Hoffman, K., Zage, D., & Nita-Rotaru, C. (2009). A survey of attack and defence techniques for reputations systems. *ACM Computing Surveys, 42*(1), 1–16. doi:10.1145/1592451.1592452

Hofstede, A. H. M. t., Proper, H. A., et al. (2007). *A note on schema equivalence*.

Holt, P., Dubs, S., Jones, M., & Greer, J. (1994). The state of student modelling. In Greer, J. E., & McCalla, G. I. (Eds.), *Student modelling: The key to individualized knowledge-based instruction* (pp. 3–35). NATO ASI, Series F., Springer-Verlag.

Hopper, M. D. (1990). Rattling sabre-new ways to compete on information. *Harvard Business Review, 68*(3), 118–125.

Horvitz, E., Breese, J., Heckerman, D., Hovel, D., & Rommelse, D. (1998). The Lurniere project: Bayesian user modeling for inferring the goals and needs of software users. In *Proceedings of the 14th Conference on Uncertainty in Artificial Intelligence,* (pp. 256- 265), Madison, WI: Morgan Kaufmann Publishers.

Hughes, B., Joshi, I., & Wareham, J. (2008). Health 2.0 and medicine 2.0: Tensions and controversies in the field. *Journal of Medical Internet Research, 10*(3), e23. doi:10.2196/jmir.1056

Hussain, S., Raza Abidi, S., & Raza Abidi, S. S. (2007). Semantic Web framework for knowledge-centric clinical decision support systems. In *Artificial Intelligence in medicine, (LNCS 4594)* (pp. 451–455). Berlin/Heidelberg, Germany: Springer. doi:10.1007/978-3-540-73599-1_60

Iribarne, L., Asensio, J. A., Padilla, N., & Ayala, R. (2008). Modelling a human-computer interaction framework for Open EMS. The Open Knowledge Society: A Computer Science and Information Systems Manifesto. In *proceedings of the 1st World Summit on the Knowledge Society* (pp. 320-327). Athens, Greece.

Isaac, A., & Summers, E. (2008). *SKOS Simple Knowledge Organization System primer: W3C working draft 29 August.* Retrieved May 22, 2010, from http://www.w3.org/TR/skos-primer/

Itmazi, J., & Megias, M. (2008). Using recommendation systems in course management systems to recommend learning objects. *The International Arab Journal of Information Technology, 5*(3), 234–240.

Jiang, P., Jones, D., & Javie, S. (2008). How third party certification programs relate to consumer trust in online transactions: An exploratory study. *Psychology and Marketing, 25*(9), 839–858. doi:10.1002/mar.20243

Juarez, J. M., Campos, M., Palma, J., & Marin, R. (2007). Computing context-dependent temporal diagnosis in complex domains. *Expert Systems with Applications, 35*(3), 991–1010. doi:10.1016/j.eswa.2007.08.054

Judson, S. R., Carver, D. L., et al. (2003). A metamodeling approach to model refactoring.

Kamel-Boulos, M. N., & Wheeler, S. (2007). The emerging Web 2.0 social software: An enabling suite of sociable technologies in health and health care education. *Health Information and Libraries Journal, 24*, 2–23. doi:10.1111/j.1471-1842.2007.00701.x

Karkaletsis, V., Stamatakis, K., Karmapyperis, P., Vojtech, M. A., Leis, A., & Villarroel, D. (2008). Automating accreditation of medical Web content. *Proceedings of the 18th European Conference on Artificial Intelligence* (ECAI 2008), 5th Prestigious Applications of Intelligent Systems (PAIS 2008), (pp. 688-692).

Kass, R. (1989). Student modelling in intelligent tutoring systems-implications for user modelling. In Kosba, A., & Wahlster, W. (Eds.), *User models in dialog systems* (pp. 386–410). Berlin, Germany: Springer-Verlag.

Kataoka, Y., & Ernst, M. D. (2001). *Automated support for program refactoring using invariants* (pp. 736–743). ICSM.

Kay, J. (1994). Lies, damned lies and stereotypes: Pragmatic approximations of users. In A. Kobsa & D. Litman (Eds.), *Proceedings of the Fourth International Conference on User Modeling* (UM'1994), (pp. 175-184). Hyannis, MA: MITRE Corp.

Kerkiri, T., Mavridis, I., & Manitsaris, A. (2009). How e-learning systems may benefit from ontologies and recommendation methods to efficiently personalize knowledge. [IJKL]. *International Journal of Knowledge and Learning*, *5*(3/4), 347–370. doi:10.1504/IJKL.2009.031229

Kerkiri, T., Paleologou, A., Konetas, D., & Hatzinikolaou, K. (2010). A learning style–driven architecture build on open source LMS's infrastructure for creation of psycho-pedagogically–savvy personalized learning paths. In Soomro, S. (Ed.), *E-learning experiences and future*.

Khan, L., & Luo, F. (2002). Ontology construction for information selection. *Proceedings of the 14th IEEE International Conference on Tools with Artificial Intelligence*, (pp. 122-131). New York, NY: IEEE Computer Society.

Kimery, K. M., & McCord, M. (2002). Third-party assurances: Mapping the road to trust

Kleppe, A., & Warmer, J. (2003). *MDA explained. The model driven architecture, practice and promise*. Addison-Wesley.

Kolbitsch, J., & Maurer, H. (2006). The transformation of the Web: How emerging communities shape the information we consume. *Journal of Universal Computer Science*, *12*(2), 187–213.

Konstantinidis, G., Flouris, G., Antoniou, G., & Christophides, V. (2007). Ontology evolution: A framework and its application to RDF. *Proceedings of the Joint ODBIS & SWDB Workshop on Semantic Web, Ontologies, Databases* (SWDB-ODBIS-07).

Kortuem, G., Kawsar, F., Sundramoorthy, V., & Fitton, D. (2010). Smart objects as building blocks for the Internet of things. *IEEE Internet Computing*, *14*(1), 44–51. doi:10.1109/MIC.2009.143

Koru, A. G., Ma, L., et al. (2003). Utilizing operational profile in refactoring large scale legacy systems. *Proceedings of the First International Workshop on REFactoring, Achievements, Challenges, Effects* (REFACE). Victoria, Canada.

Kranz, M., Holleis, P., & Schmidt, A. (2010). Embedded interaction: Interacting with the Internet of things. *IEEE Internet Computing*, *14*(2), 46–53. doi:10.1109/MIC.2009.141

Kritikou, Y., Demestichas, P., Adamopoulou, E., Demestichas, K., Theologou, M., & Paradia, M. (2008). User profile modeling in the context of Web-based learning management systems. *Science Direct. Journal of Network and Computer Applications*, *31*(4), 603–627. doi:10.1016/j.jnca.2007.11.006

Kruppa, M., Heckmann, D., & Krueger, A. (2005). Adaptive multimodal presentation of multimedia content in museum scenarios. *KI - Zeitschrift fuer Kuenstliche Intelligenz*, *5*(1), 56-59.

Langton, G. (1995). *Artificial life: An overview*. Cambridge, MA: MIT Press.

Laudon, K. C., & Laudon, J. P. (2006). *Management Information Systems: Managing the digital firm* (10th ed.). Upper Saddle River, NJ: Prentice Hall.

Leacock, C., & Chodorow, M. (1998). *Combining local context and WordNet similarity for word sense identification* (pp. 265–283).

Leslie, E., Coffee, N., Frank, L., Owen, N., Bauman, A., & Hugo, G. (2007). Walkability of local communities: Using Geographic Information Systems to objectively assess relevant environmental attributes. *Health & Place*, *13*(1), 111–122. doi:10.1016/j.healthplace.2005.11.001

Levin, S. A. (2003). Complex adaptive systems: Exploring the known, the unknown and the unknowable. *Bulletin of the American Mathematical Society*, *40*(1), 3–19. doi:10.1090/S0273-0979-02-00965-5

Lewin, R. (1999). *Complexity, life on the edge of chaos* (2nd ed.). Chicago, IL: The University of Chicago Press.

Li, M., & Vitanyi, P. (1997). *An introduction to Kolmogorov complexity and its applications*. New York, NY: Springer-Verlag.

Lieberman, H. (1995). Letizia: An agent that assists web browsing. In *Proceedings of the International Joint Conference on Artificial Intelligence,* (pp. 924–929). Montreal, Canada.

Liu, H., Lutz, C., Milicic, M., & Wolter, F. (2006). Updating description logic ABoxes. *Proceedings of the 10th International Conference on Principles of Knowledge Representation and Reasoning* (KR-06).

Lozeau, A. M., & Potter, B. (2009). Medical information and the use of emerging technologies. *Wisconsin Medical Journal, 108*(1), 30–34.

Lu, Y., Yan, Z., Laurence, T. Y., & Huansheng, N. (Eds.). (2008). *The Internet of things: From RFID to the next-generation pervasive networked systems.* Auerbach Publications, Taylor & Francis Group.

Lynch, P. D., Robert, J. K., & Srinivasan, S. S. (2001). The global Internet shopper: Evidence

Lytras, M., Tennyson, R. D., & Ordonez de Pablos, P. (2009). *Knowledge networks: The social software perspective*. Hershey, PA: IGI Global.

Lytras, M. D., & García, R. (2008). Semantic Web applications: A framework for industry and business exploitation-what is needed for the adoption of the Semantic Web from the market and industry. *International Journal of Knowledge and Learning, 4*(1), 93–108. doi:10.1504/IJKL.2008.019739

Mabbott, A., & Bull, S. (2004). Alternative views on knowledge: Presentation of open learner models. In J. C. Lester, R. M. Vicari & F. Paraguacu (Eds.), *Proceedings of the 7th International Conference on Intelligent Tutoring Systems* (ITS'2004), (pp. 689-698). Berlin/ Heidelberg, Germany: Springer-Verlag.

Maedche, A. (2002). *Ontology learning for Semantic Web*. Kluwer Academic Publishers.

Maedche, A., & Staab, S. (2001). Ontology learning for the Semantic Web. *IEEE Intelligent Systems, Special Issue on Semantic Web, 16*(2).

Maedche, A., Motik, B., Stojanovic, L., & Stojanovic, N. (2002) User-driven ontology evolution management. *Knowledge Engineering and Knowledge Management: Ontologies and the Semantic Web, 2473*(1), 133-140. Heidelberg, Germany: Springer-Verlag.

Maes, P. (1991). The agent network architecture (ANA). *SIGART Bulletin, 2*(4), 115–120. doi:10.1145/122344.122367

Magiridou, M., Sahtouris, S., Christophides, V., & Koubarakis, M. (2005). RUL: A declarative update language for RDF. *Proceedings of the 4th International Semantic Web Conference* (ISWC-05), (pp. 506-521).

Maier, R. (2007). *Knowledge management systems: Information and communication technologies for knowledge management* (3rd ed.). Berlin, Germany: Springer.

Mangiameli, P., West, D., & Rampal, R. (2004). Model selection for medical diagnosis decision support systems. *Decision Support Systems, 36*(3), 247–259. doi:10.1016/S0167-9236(02)00143-4

March, J. G. (1991). Exploration and exploitation in organizational learning. *Organization Science, 2*(1), 71–87. doi:10.1287/orsc.2.1.71

Marko Boger, T. S., & Fragemann, P. (2002). *Refactoring browser for UML*. Objects, Components, Architectures, Services, and Applications for a NetworkedWorld, International Conference NetObjectDays, NODe, 2002, (pp. 77-81).

Massa, P., & Avesani, P. (2007). Trust metrics on controversial users: Balancing between

Matsuda, N., & Okamoto, T. (1992). Student model and its recognition by hypothesis-based reasoning in ITS. *Journal of Electronics and Communications in Japan, 75*(8), 85–95. doi:10.1002/ecjc.4430750807

Mayer, R. J., Painter, M. K., & deWitte, P. S. (1992). *IDEF family of methods for concurrent engineering and business re-engineering applications*. Knowledge-Based Systems, Inc. Retrieved from http://www.idef.com/pdf/IDEFFAMI.pdf

McCalla, G. I., & Greer, J. E. (1994). Granularity-based reasoning and belief revision in student models. In *Student models: The key to individualized educational systems* (pp. 39–62). New York, NY: Springer Verlag.

McGuiness, D., Fikes, R., Rice, J., & Wilder, S. (2000). An environment for merging and testing large ontologies. *Proceedings of the 7th International Conference on Principles of Knowledge Representation and Reasoning* (KR-00), (Technical Report KSL-00-16, Knowledge Systems Laboratory, Stanford University).

McGuinness, D. L., & van Harmelen, F. (Eds.). (2004). *OWL Web ontology language overview*. Retrieved May 17, 2010, from http://www.w3.org/TR/2004/REC-owl-features-20040210/

McKeown, K., Barzilay, R., Evans, D., Hatzivassiloglou, V., Kan, M., Schiffman, B., & Teufel, S. (2001). *Columbia multi-document summarization: Approach and evaluation*. Workshop on Text Summarization.

McLean, R., Richards, B. H., & Wardman, J. I. (2007). The effect of Web 2.0 on the future of medical practice and education: Darwikinian evolution or folksonomic revolution? *The Medical Journal of Australia, 187*(3), 174–174.

McNamara, T. P., & Diwadkar, V. A. (1996). The context of memory retrieval. *Journal of Memory and Language, 35*(1), 877–892. doi:10.1006/jmla.1996.0045

Mens, T. (2004). A survey of software refactoring. *IEEE Transactions on Software Engineering, 30*(2), 126–139. doi:10.1109/TSE.2004.1265817

Mens, K., & Tourwe, T. (2004). *Delving source code with formal concept analysis*. Computer Languages, Systems & Structures.

Micarelli, A., Gasparetti, F., Sciarrone, F., & Gauch, S. (2007). Personalized search on the World Wide Web. In Brusilovsky, P., Kobsa, A., & Nejdl, W. (Eds.), *The adaptive Web: Methods and strategies of Web personalization* (pp. 195–230). Heidelberg, Germany: Springer Verlag.

Middleton, S. E., De Roure, D. C., & Shadbolt, N. R. (2001). Capturing knowledge of user preferences: Ontologies in recommender systems. In *Proceedings of the 1st International Conference on Knowledge Capture* (K-CAP '01), (pp. 100-107), Victoria, British Columbia, Canada. New York, NY: ACM Press.

Middleton, S. E., Shadbolt, N. R., & De Roure, D. C. (2003). Capturing interest through inference and visualization: ontological user profiling in recommender systems. In *Proceedings of the 2nd International Conference on Knowledge Capture* (K-CAP '03), (pp. 62-69), Sanibel Island, FL, USA. New York, NY: ACM Press.

Miller, G. A. (1995). WordNet: A lexical database for English. *Communications of the ACM, 38*(11), 39–41. doi:10.1145/219717.219748

Miller, R. A. (1994). Medical diagnostic decision support systems–past, present, and future: A threaded bibliography and brief commentary. *Journal of the American Medical Informatics Association, 1*(1), 8–27.

Missikof, M., Navigli, R., & Velardi, P. (2002). Integrated approach to Web ontology learning and engineering. *IEEE Computer, 35*(11), 60–63.

Mitchell, M. (2006). Complex systems: Network thinking. *Artificial Intelligence, 170*(18), 1194–1212. doi:10.1016/j.artint.2006.10.002

Molkenthin, R., Breuer, K., & Tennyson, R. D. (2009). Real time diagnostics of problem solving behavior for business simulations. In Baker, E., Dickieson, J., Wulfeck, W., & O'Neil, H. F. (Eds.), *Assessment of problem solving using simulations* (pp. 205–228). Mahwah, NJ: Erlbaum.

Moore, I. (1996). Automatic inheritance hierarchy restructuring and method refactoring. *Proceedings of the 11th ACM SIGPLAN conference on Object-oriented programming, systems, languages, and applications*. San Jose, California, United States, (pp. 235-250).

Mori, J., Heckmann, D., Matsuo, Y., & Jameson, A. (2007). Learning ubiquitous user models based on users' location history. *Proceedings of Data Mining for User Modeling Workshop held at the International Conference on User Modeling,* (pp. 40-49).

Muller Ross, M., Kyusuk, C., Croke, K. G., & Mensah, E. K. (2004). Geographic Information Systems in public health and medicine. *Journal of Medical Systems, 28*(3), 215–221. doi:10.1023/B:JOMS.0000032972.29060.dd

Naraine, R. (2009). 2.0 becomes security risk 2.0. In *Real business real threats*. When Web.

Nash, J. F. (1950). Equilibrium points in n-person games. *Proceedings of the National Academy of Sciences of the United States of America, 36*, 48–49. doi:10.1073/pnas.36.1.48

Neji, M. (2009). The innovative approach to user modelling for information retrieval system. *Proceedings of the 20th ACM Conference on Hypertext and Hypermedia.*

Newman, M. E. J. (2003). The structure and function of complex networks. *SIAM Review, 45*(2), 167–256. doi:10.1137/S003614450342480

NIST. (2010). *Website on summarization*. Retrieved from http://wwwnlpirnist.gov/projects/duc/pubs.html

Nonaka, I. (1991). The knowledge creating company. *Harvard Business Review, 69*, 96–104.

Nonaka, I., & Takeuchi, H. (1995). *The knowledge creating company: How Japanese companies create the dynamics of innovation*. New York, NY: Oxford University Press.

Noy, N. (2004). Semantic integration: A survey of ontology-based approaches. *SIGMOD Record, 33*(4), 65–70. doi:10.1145/1041410.1041421

Noy, N., & Klein, M. (2004). Ontology evolution: Not the same as schema evolution. *Knowledge and Information Systems, 6*(4), 428-440. (SMI technical report SMI-2002-0926).

Noy, N., & Musen, M. (2000). Algorithm and tool for automated ontology merging and alignment. *Proceedings of the 17th National Conference on Artificial Intelligence* (AAAI-00), (SMI technical report SMI-2000-0831).

Noy, N., Chugh, A., Liu, W., & Musen, M. (2006). A framework for ontology evolution in collaborative environments. *Proceedings of 5th International Semantic Web Conference*, (pp. 544–558).

Nwana, D. (1991). User modeling and user adapted interaction in an intelligent tutoring system. *User Modeling and User-Adapted Interaction, 1*, 1–32. doi:10.1007/BF00158950

O'Brien, P., & Abidi, S. (2006). Modeling intelligent ontology evolution using biological evolutionary processes. *Proceedings of IEEE International Conference on Engineering of Intelligent Systems*, (pp. 1-6).

Ohlsson, S. (1996). Learning from performance errors. *Psychological Review, 103*(2), 241–262. doi:10.1037/0033-295X.103.2.241

Oldakowsky, R., & Byzer, C. (2005). *SemMF: A framework for calculating semantic similarity of objects represented as RDF graphs*. Retrieved May 23, 2010, from http://www.corporate-semantic-web.de/pub/SemMF_ISWC2005.pdf

OMG. (2003). *OMG revised submission, UML 2.0 OCL.*

Opdyke, W. F. (1992). *Refactoring object-oriented frameworks*. Unpublished doctoral dissertation, University of Illinois.

Ostrowski, D. A. (2008). *Ontology refactoring*. Second IEEE International Conference on Semantic Computing (ICSC, 2008), (pp. 476-479).

Packer, H., Gibbins, N., & Jennings, N. (2009). Ontology evolution through agent collaboration. *Proceedings of the Workshop on Matching and Meaning 2009: Automated Development, Evolution and Interpretation of Ontologies.*

Packer, H., Gibbins, N., Payne, T., & Jennings, N. (2008) Evolving ontological knowledge bases through agent collaboration. *Proceedings of the Sixth European Workshop on Multi-Agent Systems.*

Padilla, N., Iribarne, L., Asensio, J. A., Muñoz, F., & Ayala, R. (2008). Modelling an environmental knowledge-representation system. Emerging Technologies and Information Systems for the Knowledge Society. In *proceedings of the 1st World Summit on the Knowledge Society* (pp. 70-78). Athens, Greece.

Palmer, C. L., Teffeau, L. C., & Pirmann, C. M. (2009). *Scholarly information practices in the online environment: Themes from the literature and implications for library service development.* Report commissioned by OCLC Research. Retrieved May 25, 2010, from http://www.oclc.org/programs/publications/reports/2009-02.pdf

Pan, J., Zhang, B., & Wu, G. (2009). A SAM-Based evolution model of ontological user model. *Proceedings of the 2009 Eigth IEEE/ACIS International Conference on Computer and Information Science*, (pp. 1139-1143).

Paolino, L., Sebillo, M., & Cringoli, G. (2005). Geographical Information Systems and online GIServices for health data sharing and management. *Parassitologia, 47*(1), 171–175.

Parra, D., & Brusilovsky, P. (2009). Collaborative filtering for social tagging systems: An experiment with CiteULike. Proceedings of the Third ACM Conference on Recommender Systems, New York, USA, (pp. 237-240).

Pavard, B., & Dugdale, J. (2000). *The contribution of complexity theory to the study of sociotechnical cooperative systems.* Paper presented at the Third International Conference on Complex Systems, Nashua, NH, May 21-26, 2000. Retrieved from http://www-svcict.fr/cotcos/pjs/

Pazzani, M. J., & Billsus, D. (2007). Content-based recommendation systems. In Brusilovsky, P., Kobsa, A., & Nejdl, W. (Eds.), *The adaptive Web: Methods and strategies of Web personalization* (pp. 325–341). Heidelberg, Germany: Springer Verlag.

Pedraza-Jiménez, R., Valverde-Albacete, F., & Navia-Vázquez, A. (2006). A generalisation of fuzzy concept lattices for the analysis of Web retrieval tasks. *Proceedings of the 6th International Conference on Information Processing and Management of Uncertainty in Knowledge-Based Systems*, Paris, France. New York, NY: IEEE

Peng, Z. R., & Tsou, M. H. (2003). *Internet GIS –distributed GIServices for the Internet and wireless networks.* Wiley.

Pereira, C., & Tettamanzi, A. (2006). An evolutionary approach to ontology-based user model acquisition. *Fuzzy Logic and Applications, 2955*(1), 25-32. Heidelberg, Germany: Springer Verlag.

Pérez-Marín, D., Alfonseca, E., & Pascual-Nieto, I. (2007). Automatic generation of students' conceptual models from answers in plain text. In C. Conati, K. McCoy & G. Paliouras (Eds.), *Proceedings of the 11 International Conference on User Modeling* (UM'2007), (pp. 329-333). Corfu, Greece. Berlin / Heidelberg, Germany: Springer.

Philipps, J., & Rumpe, B. (2001). *Roots of refactoring.* Tenth OOPSLA Workshop on Behavioral Semantics. Tampa Bay, Florida, USA.

Pierrakos, D., Paliouras, G., Papatheodorou, C., & Spyropoulos, C. D. (2003). Web usage mining as a tool for personalization: A survey. *User Modeling and User-Adapted Interaction, 13*(4), 311–372. doi:10.1023/A:1026238916441

Plessers, P., & de Troyer, O. (2005). Ontology change detection using a version log. *Proceedings of the 4th International Semantic Web Conference* (ISWC-05), (pp. 578-592).

Podgorelec, V., Grasic, B., & Pavlic, L. (2009). Medical diagnostic process optimization through the semantic integration of data resources. *Computer Methods and Programs in Biomedicine, 95*(2), S55–S67. doi:10.1016/j.cmpb.2009.02.015

Polat, K., & Güneş, S. (2007). An expert system approach based on principal component analysis and adaptive neuro-fuzzy inference system to diagnosis of diabetes disease. *Digital Signal Processing, 17*(4), 702–710. doi:10.1016/j.dsp.2006.09.005

Popescul, A., Ungar, L. H., Pennock, D. M., & Lawrence, S. (2001). Probabilistic models for unified-collaborative and content-based recommendation in sparse-data environments. In J. S. Breese & D. Koller (Eds.), *Proceedings of the 17th Conference on Uncertainty in Artificial Intelligence (UAI)*, (pp. 437-44). Massachusetts: Morgan Kaufmann.

Porres, I. (2002). *A toolkit for manipulating UML models.* TUCS Turku Centre for Computer Science.

Porres, I. (2003). *Model refactorings as rule-based update transformations. UML, 2003* (pp. 159–174). Springer.

Proper, H. A., & Halpin, T. A. (2004). *Conceptual schema optimisation-database optimisation before sliding down the waterfall* (p. 34). Department of Computer Science - University of Queensland.

Putnam, R. D. (2000). *Bowling alone.* New York, NY: Simon & Schuster.

Qi, Z., Zheng, Y., & Jiang, X. (2008). An approach for domain ontology-based semantic retrieval. In *proceedings of 3rd China-Ireland International Conference on Information and Communications Technologies* (pp. 185-189). Beijing.

Ragnemalm, E. L. (1999). *Student modelling based on collaborative dialogue with a learning companion.* Dissertation No 563, Linkoping University, S-581 83, Linkoping, Sweden.

Rahman, A., Alam, H., Hartono, R., & Ariyoshi, K. (2001). *Automatic summarization of Web content to smaller display devices*. 6th International Conference on Document Analysis and Recognition, ICDAR01, (pp. 1064-1068).

Rajapakse, M., Kanagasabai, R., Ang, W. T., Veeramani, A., Schreiber, M. J., & Baker, C. J. O. (2008). Ontology-centric integration and navigation of the Dengue literature. *Journal of Biomedical Informatics*, *41*(5), 806–815. doi:10.1016/j.jbi.2008.04.004

Razmerita, L., & Gouardères, G. (2004). Ontology based user modeling for personalization of Grid learning services. *Proceedings of the Grid Learning Services Workshop (GLS 2004) in Association with Intelligent Tutoring System Conference*.

Reffay, C., & Chanier, T. (2002). *Social network analysis used for modelling collaboration in distance learning groups*. (LNCS 2363), (pp. 31-40).

Resnick, P., & Varian, H. (1997). Recommender systems, introduction to special section. *Communications of the ACM*, *40*(3).

Resnick, P., Zeckhauser, R., Friedman, E., & Kuwabara, K. (2000). Reputation systems: Facilitating trust in Internet interactions. *Communications of the ACM*, *43*(12), 45–48. doi:10.1145/355112.355122

Resnik, P. (1995). *Disambiguating noun groupings with respect to WordNet senses*. Chelmsford, MA: Sun Microsystems Laboratories.

Roberts, D. B. (1999). *Practical analysis for refactoring*. Unpublished doctoral dissertation, University of Illinois.

Roberts, D., Brant, J., et al. (1997). A refactoring tool for smalltalk. *Theory and Practice of Object Systems, 3*(4).

Rodrigo, M., Baker, R., Maria, L., Sheryl, L., Alexis, M., Sheila, P., et al. (2007). Affect and usage choices in simulation problem solving environments. In R. Luckin, K. R. Koedinger & J. Greer (Eds.), *Proceedings of the 13th International Conference on Artificial Intelligence in Education* (AIED'2007), (pp. 145-152), Marina Del Ray, CA, USA. IOS Press.

Roger, M., Simonet, A., & Simonet, M. (2002). Toward updates in description logics. *Proceedings of the 2002 International Workshop on Description Logics* (DL-02), CEUR-WS 53(1).

Romano, A., De Maggio, M., & Del Vecchio, P. (2009). The emergence of a new managerial mindset. In Romano, A. (Ed.), *Open business innovation leadership. The emergence of the stakeholder university* (pp. 19–65). UK: Palgrave Macmillan.

Rovai, A. P., & Barnum, K. T. (2003). Online course effectiveness: An analysis of student interactions and perceptions of learning. *Journal of Distance Education*, *18*, 57–73.

Rysselberghe, F. V., & Demeyer, S. (2004). Evaluating clone detection techniques from a refactoring perspective. *Proceedings of the 19th International Conference on Automated Software Engineering* (ASE'04). Linz, Austria, (pp. 336-339).

Sacco, G. (2000). Dynamic taxonomies, a model for large information bases. *IEEE Transactions on Data and Knowledge Engineering*, *12*(3), 468–479. doi:10.1109/69.846296

Salton, G., Wong, A., & Yang, C.-S. (1975). A vector space model for automatic indexing. *Communications of the ACM*, *18*(11), 613–620. doi:10.1145/361219.361220

Salton, G. (1971). The Smart retrieval system–experiments. In Salton, G. (Ed.), *Automatic document processing*. New Jersey: Prentice–Hall.

Sandars, J., & Schroter, S. (2007). Web 2.0 technologies for undergraduate and postgraduate medical education: An online survey. *Postgraduate Medical Journal*, *83*, 759–762. doi:10.1136/pgmj.2007.063123

Sarwar, B., Karypis, G., Konstan, J., & Riedl, J. (2000). Analysis of recommendation algorithms for e-commerce. In A. Jhingran, J. M. Mason & D. Tygar (Eds.), *Proceedings of ACM E-Commerce 2000 conference*, (pp. 158-167). New York, NY: ACM.

Schafer, J., Frankowski, D., Herlocker, J., & Sen, S. (2007). Collaborative filtering recommender systems. In Brusilovsky, P., Kobsa, A., & Nejdl, W. (Eds.), *The adaptive Web* (pp. 291–324). Berlin/ Heidelberg, Germany: Springer-Verlag. doi:10.1007/978-3-540-72079-9_9

Schein, A. I., Popescul, A., & Ungar, L. H. (2002). Methods and metrics for cold-start recommendations. In K. Jarvelin, M. Beaulieu, R. Baeza-Yates & S. H. Myaeng, (Eds), *Proceedings of the 25'th Annual International ACM SIGIR Conference on Research and Developmentin Information Retrieval (SIGIR 2002)*, (pp. 253-260). New York, NY: ACM.

Schilit, B. N., & Heimer, M. M. (1994). Disseminating active map information to mobile hosts. *IEEE Network*, *8*(5), 22–32. doi:10.1109/65.313011

Schiller, J., & Voisard, A. (2004). *Location-based services*. Amsterdam, The Netherlands: Elsevier.

Schleyer, T., Spallek, H., Butler, B. S., Subramanian, S., Weiss, D., & Poythress, M. S. (2008). Facebook for scientists: Requirements and services for optimizing how scientific collaborations are established. *Journal of Medical Internet Research*, *10*(3), e24. doi:10.2196/jmir.1047

Schott, F., & Hillebrandt, D. (Eds.). (2005). *Outside behavior–inside cognition?*Berlin, Germany: Springer.

Schulz, S., Stenzhorn, H., Boeker, M., & Smith, B. (2009). Strengths and limitations of formal ontologies in the biomedical domain. *Electronic Journal of Communication Information & Innovation in Heath*, *3*(1), 31–45.

Sciulli, L. M. (1998). How organizational structure influences success in various types of innovation. *Journal of Retail Banking Services*, *20*(1), 13–18.

Sedayao, J., Su, S., Ma, X., Jiang, M., & Miao, K. (2009). *A simple technique for securing data at rest stored in a computing cloud.* (LNCS 5931), (pp. 553-558).

Seeman, N. (2008). Web 2.0 and chronic illness: New horizons, new opportunities. *Healthcare Quarterly (Toronto, Ont.)*, *11*(1), 104–110.

Self, J. (1994). Formal approaches to student modeling. In McCalla, G., & Greer, J. (Eds.), *Student models: The key to individualized educational systems*. New York, NY: Springer Verlag.

Shahar, Y., & Cheng, C. (1998). *Knowledge-based visualization of time oriented clinical data*. AMIA Annual Fall Symposium, (pp. 155-9).

Sheng, O. R. L. (2004). Editorial: Decision support for healthcare in a new information age. *Decision Support Systems*, *30*(2), 101–103. doi:10.1016/S0167-9236(00)00091-9

Sicilia, M.-A., Garcia, E., Diaz, P., & Aedo, I. (2004). Using links to describe imprecise relationships in educational contents. *International Journal of Continuing Engineering Education and Lifelong Learning*, *14*(3), 260–275. doi:10.1504/IJCEELL.2004.004973

Sicilia, J. J., Sicilia, M. A., Sánchez-Alonso, S., García-Barriocanal, E., & Pontikaki, M. (2009). Knowledge representation issues in ontology-based clinical knowledge management systems. *International Journal of Technology Management, 47*(1/2/3), 191-206.

Simmonds, J., & Mens, T. (2002). *A comparison of software refactoring tools*. Programing Tecnology Lab.

Simon, F., Steinbrukner, F., et al. (2001). Metrics based refactoring. *Proceedings of the 5th European Conference on Software Maintenance and Reengineering*, IEEE Computer Society Press.

Simonet, M., Messai, R., Diallo, G., & Simonet, A. (2008). Ontologies in the health field. In Berka, P., Rauch, J., & Abdelkader Zighed, D. (Eds.), *Data mining and medical knowledge management: Cases and applications*. Hershey, PA: IGI Global.

Sison, R., & Shimura, M. (1998). Student modeling and machine learning international. *Journal of Artificial Intelligence in Education*, *9*, 128–158.

Sivadas, E., & Dwyer, F. R. (2000). An examination of organizational factors influencing new product success in internal and alliance based processes. *Journal of Marketing*, *64*(1), 31–50. doi:10.1509/jmkg.64.1.31.17985

Sora, I., Todinca, D., & Avram, C. (2009). *Translating user preferences into fuzzy rules for the automatic selection of services*. In the 5th International Symposium on Applied Computational Intelligence and Informatics, IEEE. (pp. 487-492). Timisoara, Romania.

Sosnovsky, S., & Dicheva, D. (2010). Ontological technologies for user modeling. *International Journal of Metadata. Semantics and Ontologies*, *5*(1), 32–71. doi:10.1504/IJMSO.2010.032649

Sousan, W. L., Payne, M., Nickell, R., & Zhu, Q. (2007). *Metadata (ontology) incremental building and refinement agents*. In 2007 International Conference on Integration of Knowledge Intensive Multi-Agent Systems, IEEE. (pp. 127-132). Waltham.

Souza, J. F., Paula, M., Oliveira, J., & Souza, J. M. (2006). *Meaning negotiation: Applying negotiation models to reach semantic consensus in multidisciplinary teams. Proceedings of Group Decision and Negotiation* (pp. 297–300). Karlsruhe: Universitätsverlag Karlsruhe.

Souza, J. F., Melo, R. N., Oliveira, J., & Souza, J. M. (2009b). Combining resemblance functions for ontology alignment. *Proceedings of the 11th International Conference on Information Integration and Web-based Applications & Services* (iiWAS2009).

Souza, J. F., Siqueira, S. W. M., & Melo, R. N. (2009a). Adding meaning negotiation skills in multiagent systems. *Proceedings of IEEE International Conference on Intelligent Computing and Intelligent Systems*, (pp. 663-667). Shangai, China: IEEE Press.

Spector, J. M., Ohrazda, C., & Van Schaak, A. (Eds.). (2005). *Innovations in instructional technology: Essays in honor of M. David Merrill*. Mahwah, NJ: Erlbaum.

Spiess, P., Karnouskos, S., Guinard, D., Savio, D., Baecker, O., Souza, L. M., & Trifa, V. (2009). *SOA-based integration of the Internet of things in enterprise services*. IEEE International Conference on Web Services, ICWS 2009, (pp. 968-975).

Stahl, C., Heckmann, D., Schwartz, T., & Fickert, O. (2007). Here and now: A user-adaptive and location-aware task planner. In S. Berkovsky, K. Cheverst, P. Dolog, D. Heckmann, T. Kuflik, P. Mylonas, ... J. Vassileva (Eds.), *Proceedings of the Workshop on Ubiquitous and Decentralized User Modeling (UbiDeUM) at UM'200,7* (pp. 52-67).

Steels, L. (1995). When are robots intelligent autonomous agents. *Robotics and Autonomous Systems*.

Stewart, A., Niederée, C., & Mehta, B. (2004). *State of the art in user modelling for personalization in content, service and interaction*. NSF/DELOS Report on Personalization.

Stinchcombe, A. L. (1990). *Information and organizations*. Berkley, CA: University of California Press.

Stojanovic, L., Maedche, A., Motik, B., & Stojanovic, N. (2002). User-driven ontology evolution management. *Proceedings of the 13th International Conference on Knowledge Engineering and Knowledge Management* (EKAW-02), (LNCS 2473), (pp. 285-300). Heidelberg, Germany: Springer-Verlag.

Stompka, P. (1999). *Trust*. Cambridge, UK: Cambridge University Press.

Stuckenschmidt, H., & Klein, M. (2003). Integrity and change in modular ontologies. *Proceedings of the 18th International Joint Conference on Artificial Intelligence* (IJCAI-03). Jaziri W., Sassi N., & Gargouri F. (2010). Approach and tool to evolve ontology and maintain its coherence. *International Journal of Metadata, Semantics and Ontologies.* United Kingdom: Inderscience Publishers.

Studer, R., Benjamins, V. R., & Fensel, D. (1998). Knowledge engineering: Principles and methods. *Data & Knowledge Engineering*, *25*(1-2), 161–197. doi:10.1016/S0169-023X(97)00056-6

Subasi, A. (2007). Application of adaptive neuro-fuzzy inference system for epileptic seizure detection using wavelet feature extraction. *Computers in Biology and Medicine*, *37*(2), 227–244. doi:10.1016/j.compbiomed.2005.12.003

Sunye, G., & Pollet, D. (2001). Refactoring UML models. In Gogolla, M., & Kobryn, C. (Eds.), *UML* (pp. 134–148). Springer.

Sváb-Zamazal, O., Svátek, V., Meilicke, C., & Stuckenschmidt, H. (2008). *Testing the impact of pattern-based ontology refactoring on ontology matching results*. 3rd International Workshop on Ontology Matching.

Svozil, K., & Wright, R. (2005). Statistical structures underlying quantum mechanics and social science. *International Journal of Theoretical Physics*, *44*(7), 1067–1086.

Symantec Enterprise Security. (2009). *Internet security threat report* (*Vol. XIV*). Symantec.

Tahvildari, L. (2003). *Quality-driven object-oriented code restructuring*. In the IEEE Software Technology and Engineering Practice (STEP) - Workshop on Software Analysis and Maintenance, Practices, Tools, Interoperability (SAM). Amsterdam, The Netherlands.

Taifi, N., Corallo, A., & Passiante, G. (2008). The strategic orientation of the managerial ties for the new product development. *International Journal of Knowledge and Learning, 4*(6), 613–624. doi:10.1504/IJKL.2008.022892

Tan, T. Z., Quek, C., Ng, G. S., & Razvi, K. (2008). Ovarian cancer diagnosis with complementary learning fuzzy neural network. *Artificial Intelligence in Medicine, 43*(3), 207–222. doi:10.1016/j.artmed.2008.04.003

Taniar, D., & Rahayu, J. (2004). *Web Information Systems*. Hershey, PA: IGI Global.

Tennyson, R. D., & Jorczak, R. L. (2008). A conceptual framework for the empirical study of games. In O'Neil, H., & Perez, R. (Eds.), *Computer games and team and individual learning* (pp. 3–20). Mahwah, NJ: Erlbaum.

Tennyson, R. D., Wu, J.-H., & Hsia, T.-L. (in press). Engaging learning models with information and communication technologies in advancing electronic learning. In Ordóñez de Pablos, P. (Ed.), *Electronic globalized business and sustainable development through IT management: Strategies and perspectives*. Hershey, PA: IGI Global.

Thagard, P. (1992). *Conceptual revolutions*. Princeton University Press.

Thomas, J. P., Wiechert, F., Florian, M., Patrick, S., & Elgar, F. (2010). *Connecting mobile phones to the Internet of things: A discussion of compatibility issues between EPC and NFC*. Auto-ID Labs.

Thor, A., et al. (2009). An evolution-based approach for assessing ontology mappings-a case study in the life sciences. *Proceedings of the 13th Conf. of Database Systems for Business, Technology and Web* (BTW).

Tourwe, T., & Mens, T. (2003). Identifying refactoring opportunities using logic meta programming. *Proceedings of the 7th European Conference on Software Maintenance and Reengineering*, IEEE Computer Society Press.

Trader, I. (1996). *ISO/IEC DIS 13235-1: IT-ODP trading function-part 1: Specification.*

Tripathi, A., Suman Kumar Reddy, T., Madria, S., & Mohanty, H. (2009). Algorithms for validating e-tickets in mobile computing environment. *Information Sciences, 179*(11), 1678–1693. doi:10.1016/j.ins.2009.01.018

Tsai, K.-H., Chiu, T. K., Ming, C.-L., & Wang, T.-I. (2006). A learning objects recommendation model based on the preference and ontological approaches. *Proceedings of the Sixth International Conference on Advanced Learning Technologies* (ICALT'06), (pp. 36-40).

Übeyli, E. D. (2007). Implementing automated diagnostic systems for breast cancer detection. *Expert Systems with Applications, 33*(4), 1054–1062. doi:10.1016/j.eswa.2006.08.005

Ueno, M. (2003). *LMS with irregular learning processes detection system*. In World Conference on e-Learning in Corp., Govt., Health, & Higher Education. (pp. 2486-2493).

Upendra, S., & Patti, M. (1995). Social information filtering: Algorithms for automating word of mouth. *Proceedings of the SIGCHI conference on Human factors in computing systems*, (pp. 210-217).

Urban, G. L., & Sultan, F. (2000). Placing trust at the center of your Internet strategy.

User Interface Android. (2010). *Specifications*. Retrieved from http://developer.android.com/guide/topics/ui/index.html

Van Bemmel, J. H., & Musen, M. A. (1997). *Handbook of medical informatics*. Springer-Verlag.

van Rijsbergen, C. J. (1979). *Information retrieval*. London, UK: Butterworths.

Van Wyck, K. (2009). How to use Facebook safely. *IT manager's guide to social networking.*

VanLehn, K. (1988). Student modeling. In Polson, M. C., & Richardson, J. J. (Eds.), *Foundations of intelligent tutoring systems* (pp. 55–78).

VanLehn, K., & Martin, J. (1997). Evaluation of an assessment system based on Bayesian student modelling, 8. (pp. 179-221).

Varela, F. J., & Bourgine, P. (1992). Toward a practice of autonomous systems. *Proceedings of the First European Conference on Artificial Life*. Cambridge, MA: MIT Press, Bradford Books.

Von Neumann, J., & Mongerstern, O. (1953). *Theory of games and econonmic behavior* (3rd ed.). Princeton, NJ: Princeton University Press.

Vossen, G., Lytras, M., & Koudas, N. (2007). Editorial: Revisiting the (machine) Semantic Web: The missing layers for the human Semantic Web. *IEEE Transactions on Knowledge and Data Engineering*, *19*(2), 145–148. doi:10.1109/TKDE.2007.30

Wagner, E. D. (1994). In support of a functional definition of interaction. *American Journal of Distance Education*, *8*(2), 6–26. doi:10.1080/08923649409526852

Wang, F., & Luo, W. (2005). Assessing spatial and non-spatial factors for healthcare access: Towards an integrated approach to defining health professional shortage areas. *Health & Place*, *11*(2), 131–146. doi:10.1016/j.healthplace.2004.02.003

Wang, T. I., Tsai, K. H., Lee, M. C., & Chiu, T. K. (2007). Personalized learning objects recommendation based on the semantic-aware discovery and the learner preference pattern. *Journal of Educational Technology & Society*, *10*(3), 84–105.

Wang, Y. D., & Emurian, H. (2005). An overview of online trust: Concepts, elements, and

Wasserman, S., & Faust, K. (1996). *Social network analysis. Methods and applications*. Cambridge, MA: Cambridge University Press.

Webb, G. I., Pazzani, M. J., & Billsus, D. (2001). Machine learning for user modeling. *International Journal of User Modeling and User-Adapted Interaction*, *11*(1-2), 19–29. doi:10.1023/A:1011117102175

Wheelwright, S. C., & Clark, K. B. (1992). *Revolutionizing product development*. New York, NY: Free Press.

WHO. (2009). *ICD: International statistical classification of diseases and related health problems*. World Health Organization. Retrieved from http://www.who.int/classifications/icd/en/

Wiig, K. M. (1995). *Knowledge management foundations-thinking about thinking-how people and organizations create, represent, and use knowledge*. Arlington, TX: Schema Press.

Wikipedia. (2010). *Collaborative filtering*. Retrieved from http://en.wikipedia.org/wiki/Collaborative_filtering

Wikipedia. (2010). *Reputation system*. Retrieved from http://en.wikipedia.org/wiki/Reputation_ system

Wikipedia. (2010). *Social software*. Retrieved from http://en.wikipedia.org/wiki/Social_software

Wordnet. (2009). *Homepage*. Retrieved May 07, 2010, from http://wordnet.princeton.edu/

Wu, J.-H., Hsia, T.-L., & Tennyson, R. D. (in press). Design strategies for improved online instructional systems. In Ordóñez de Pablos, P. (Ed.), *Electronic globalized business and sustainable development through IT management: Strategies and perspectives*. Hershey, PA: IGI Global.

Xiao-Feng, M., Bao-Wen, X., Qing, L., Ge, Y., Jun-Yi, S., Zheng-Ding, L., & Yan-Xiang, H. (2006). A survey of Web Information Technology and application. [Wuhan University.]. *Journal of Natural Sciences*, *11*(1), 1–5.

Xing, Z., & Stroulia, E. (2003). Recognizing refactoring from change tree. *Proceedings of the First International Workshop on REFactoring, Achievements, Challenges, Effects* (REFACE). Victoria, Canada.

Yager, R. R. (2007). Centered OWA operators. *Soft Computing*, *11*(7), 632–639. doi:10.1007/s00500-006-0125-z

Yan, H., Zheng, J., Jiang, Y., Peng, C., & Xiao, S. (2008). Selecting critical clinical features for heart diseases diagnosis with a real-coded genetic algorithm. *Applied Soft Computing*, *8*(2), 1105–1111. doi:10.1016/j.asoc.2007.05.017

Yeon-Seok, K., & Kyong-Ho, L. (2007). A light-weight framework for hosting Web services on mobile devices. *Proceedings of the 5th IEEE European Conference on Web Services*, ECOWS 07, (pp. 255-263).

Yong Rui, Y., Gupta, A., & Acero, A. (2000). Automatically extracting highlights for TV baseball programs. *ACM Multimedia*, 105-115.

Yousafzai, S., Pallister, J., & Foxall, G. (2005). E-banking-a matter of trust: Trust-

Yu, W., Li, L., et al. (2004). Refactoring use case models on episodes. *Proceedings of the 19th International Conference on Automated Software Engineering* (ASE'04). Linz, Austria, (pp. 328-331).

Yu, Y., Mylopoulos, J., et al. (2003). Software refactoring guided by multiple soft-goals. *Proceedings of the First International Workshop on REFactoring, Achievements, Challenges, Effects* (REFACE). Victoria, Canada.

Zablith, F. (2008). Dynamic ontology evolution. *Proceedings of the International Semantic Web Conference (ISWC)* Doctoral Consortium. Karlsruhe, Germany.

Zablith, F. (2009). Ontology evolution: A practical approach. *Proceedings of the AISB Workshop on Meaning and Matching* (WMM), Edinburgh, UK.

Zadeh, L. A. (2003). The concept of a linguistic variable and its applications to approximate reasoning. *Information Sciences*, *9*(3), 43–80.

Zein, O. K., Kermarrec, Y., & Salaun, M. (2006). An approach for discovering and indexing services for self-management in autonomic computing systems. *Annales des Télécommunications*, *61*(9-10), 1046–1065.

Zhang, J., & Lin, Y. (2005). Generic and domain-specific model refactoring using a model transformation engine. In *Model-driven software development-research and practice in software engineering*. Springer. doi:10.1007/3-540-28554-7_9

Zhang, H., Song, Y., & Song, H.-t. (2007). Construction of ontology-based user model for Web Personalization. In C. Conati, K. McCoy & G. Paliouras (Eds.), *Proceedings of the 11 International Conference on User Modeling* (UM'2007), (pp. 67-76), Corfu, Greece. Berlin /Heidelberg, Germany: Springer.

Zhou, L., & Hovy, E. (2006). *On the summarization of dynamically introduced information: Online discussions and blogs*. In AAAI Spring Symposium on Computational Approaches to Analysing Weblogs.

Ziegler, C. (2004). Semantic Web recommender systems. In Lindner, W. (Ed.), *EDBT 2004 Workshops* (pp. 78–89). Berlin/Heidelberg, Germany: Springer-Verlag.

Ziegler, C., & Golbeck, J. (2006). Investigating correlations of trust and interest similarity.

Zimmer, J. A. (2003). Graph theoretical indicators and refactoring. In F. M. a. D. Wells (Ed.), *Proceedings of the Third XP and Segond Agile Universe Conference.* New Orleans, USA, Springer. (pp. 27-53).

Zirkin, B., & Sumler, D. (1995). Interactive or non-interactive? That is the question! An annotated bibliography. *Journal of Distance Education*, *10*(1), 95–112.

About the Contributors

Miltiadis D. Lytras is a professor in the American College of Greece. His research focuses on Semantic Web, knowledge management and e-learning, with more than 100 publications in these areas. He has co-edited /co-edits, 25 special issues in International Journals (e.g. IEEE Transaction on Knowledge and Data Engineering, IEEE Internet Computing, IEEE Transactions on Education, Computers in Human Behaviour, Interactive Learning Environments, Journal of Knowledge Management, Journal of Computer Assisted Learning, etc) and has authored/[co-]edited 25 books.

Patricia Ordóñez de Pablos is a Professor in the Department of Business Administration and Accountability in the Faculty of Economics of the University of Oviedo, Spain. Her teaching and research interests focus on the areas of strategic management, knowledge management, intellectual capital measuring and reporting, and Information Technologies. She serves as Editor in Chief of "International Journal of Asian Business and Information Management" (IGI-Global) and "International Journal of Chinese Culture and Management". She also serves as Associate Editor of "Behaviour and Information Technology" (Francis & Taylor). Additionally she serves as Executive Editor of "International Journal of Learning and Intellectual", "International Journal of Strategic Change Management", and "International Journal of Arab Culture, Management and Sustainable Development", respectively. She is co-editor of "Advances in Strategic Management and Sustainable Development" Series (IGI-Global) and "Advances in Emerging Information Technology Issues" Series (IGI-Global). She has co-edited/co-edits 20 special issues in IS Thompson International Journals (such as International Journal of Technology Management; Interactive Leaning Environments; Ergonomics and Human Factors in Manufacturing; Information Systems Management, among others). She has co-edited more 10 books, such as The China IT Handbook (Springer); Web 2.0: The Business Model (Springer); Technology Enhanced Learning: Best Practices (IGI-Global); Knowledge Ecology in Global Business: Managing Intellectual Capital (IGI Global); Social Web Evolution: Integrating Semantic Applications and Web 2.0 Technologies (IGI-Global); Emerging Topics And Technologies In Information Systems (IGI Global); Knowledge Networks: The social software perspective (IGI Global); Information Technology, Information Systems and Knowledge Management (Springer).

Ernesto Damiani is currently a professor at the Università degli Studi di Milano and the director of the Università degli Studi di Milano's PhD program in computer science. He has held visiting positions at a number of international institutions, including George Mason University in Virginia, LaTrobe University in Melbourne, Australia, University of Technology in Sydney, Australia and the Institut National des Sciences Appliquées (INSA) at Lyon, France.

* * *

José A. Asensio is a doctoral student at the University of Almeria (UAL), Spain. He received the BSc and MSc degrees in Computer Science from the UAL. Since 2006, he works in two national research projects on Environmental Management Systems. In 2007, he joined Applied Computing Group and then became Associate Professor, University of Almería in 2009. Since 2003, he works as a Technician in the Unit of Support Technologies for Learning and Virtual Learning at the University of Almería. His research interests include trading, agents & multi-agent systems (MAS), ontology-driven engineering (ODE), model-driven engineering (MDE), model transformations, and system modelling.

Cristian Bisconti achieved his PhD in Nuclear Systems in February 2006 at the University of Salento, and was rewarded with the national award Fubini for the best PhD thesis in Theoretical Atomic Physics. He carried on his studies in nuclear physics in collaboration with the Department of Atomic and Nuclear Physics at University of Granada (Spain) and the Department of Physics E. Fermi of Pisa. Since March 2008, he started working at Scuola Superiore ISUFI, University of Salento focusing on complex systems in techno-economic contexts. His research interest is in applying methodologies and tools of quantum physics to social sciences like the dynamics of inter-organizational structures and informal communities, with specific attention to the relationship between technology and organization.

Vladlena Benson, is Senior Lecturer at the Faculty of Business and Law, Kingston University, UK. Vladlena's research interests include information management, e-learning and Web technologies, including social networking. She has published a number of papers, books, and invited chapters in the area of Information Technology. Vladlena teaches managing information, Information Systems development and information security on postgraduate and undergraduate programmes.

Santi Caballé received his PhD, Masters and Bachelors in Computer Science from the Open University of Catalonia (Barcelona, Spain). Since 2003, he has been an Assistant Professor at the Open University of Catalonia, where he became Associate Professor in 2006 teaching a variety of online courses in Computer Science in the area of Software Engineering, Information Systems and Collaborative Learning. Dr. Caballé has been involved in the organization of several international conferences, conference tracks and workshops, and has published over 60 research contributions as books, book chapters, and referred international journal and conference papers. He has also acted as editor for books and special issues of international journals. His research focuses on software engineering, e-learning and collaborative and mobile learning, and distributed, Grid and peer-to-peer technologies.

Eliana Campi has a degree in computer science engineering with a thesis on the modeling and implementation of ontology to support the local development. She is currently a research fellow at the Center for Business Innovation, University of Salento, Italy. Her expertise regards the ontology application for knowledge management, the innovative technologies and methodologies to organize knowledge bases, to index and to search information based on metadata, and the innovative approaches based on the vision of the Semantic Web.

Antonella Carbonaro is an Associate Professor of Computer Science at the Department of Computer Science of the University of Bologna. She received the Italian Laurea degree in Computer Science in 1992. In 1997 she finished her Ph.D. studies at University of Ancona on Artificial Intelligent System. From 1997 to 1999, she received a research fellowship with theme of search artificial intelligence. Her current research interests concern personalization and content-based services for data and knowledge mining and personalized learning environment.

José Manuel Morales del Castillo is MLS graduate and PhD with the Information Science Department at the University of Granada. His papers have been published in different journals specialized on library and Information Science, computational intelligence, fuzzy logic and soft computing. His research lines are focused on the application of Semantic Web technologies to recommender systems, digital libraries and e-learning systems in order to improve information representation and processing.

Valerio Cisternino has a degree in computer science engineering with a thesis on the development of an environment enabling knowledge codification processes and implementing an ontology reasoned. He is a research fellow at the Center for Business Innovation, University of Salento, Italy. He participated to the development of object-oriented software systems based on open Internet standards and J2EE technologies. He is currently involved in the new product design area related to the customization of PDM systems, the design of an infrastructure for a machine workshop integration system and the modeling of business processes.

Ricardo Colomo-Palacios is an Associate Professor at the Computer Science Department of the Universidad Carlos III de Madrid. His research interests include applied research in Information Systems, software project management, people in software projects and social and Semantic Web. He received his PhD in Computer Science from the Universidad Politécnica of Madrid (2005). He also holds a MBA from the Instituto de Empresa (2002). He has been working as software engineer, project manager and software engineering consultant in several companies, including Spanish IT leader, INDRA. He is also an Editorial Board Member and Associate Editor for several international journals and conferences and Editor in Chief of International Journal of Human Capital and Information Technology Professionals.

Jordi Conesa is an assistant professor of Information Systems at Universitat Oberta de Catalunya, Barcelona, Spain. His research interest concerns the areas of conceptual modeling, ontologies, Semantic Web, knowledge-based systems and e-learning. His long-term goal is to develop methodologies and tools to use ontologies effectively in several application domains, such as conceptual modelling, software engineering and e-learning. He received his M.S. and Ph.D. in Software Engineering from the Technical University of Catalonia.

Angelo Corallo is assistant professor at the Center for Business Innovation and researcher at the Department of Innovation Engineering, University of Salento, Italy. His research interests encompass mainly the co-evolution of the technological and organizational innovation, with a specific focus on business ecosystems, the extended enterprise, knowledge management and collaborative working environments in project-based organizations. He is currently involved in many research projects related to technological and organizational issues both at Italian and European levels. He collaborates with Massachusetts Institute of Technology (MIT) in projects related to enterprise process modeling and social network analysis.

Angel García-Crespo is the Head of the SofLab Group at the Computer Science Department in the Universidad Carlos III de Madrid and the Head of the Institute for Promotion of Innovation Pedro Juan de Lastanosa. He holds a PhD in Industrial Engineering from the Universidad Politécnica de Madrid (Award from the Instituto J.A. Artigas to the best thesis) and received an Executive MBA from the Instituto de Empresa. Professor García-Crespo has led and actively contributed to large European projects of the FP V and VI, and also in many business cooperations. He is the author of more than a hundred publications in conferences, journals and books, both Spanish and international.

Javier Criado is a doctoral student at the University of Almeria (UAL), Spain. He received the BSc and MSc degrees in Computer Science from the UAL. Since 2009, he works as a researcher in the project SOLERES of the Spanish Ministry of Science and Innovation (MICINN). In 2009, he joined Applied Computing Group. His research interests include model-driven engineering (MDE), model transformations, ontology-driven engineering (ODE), human-computer interaction (HCI) for advances on user interfaces, COTS components, trading, agents & MAS, and UML design.

Juan Manuel Cueva Lovelle is a Mining Engineer from Oviedo Mining Engineers Technical School in 1983 (Oviedo University, Spain). Ph. D. from Madrid Polytechnic University, Spain (1990). From 1985 he is a Professor at the Languages and Computers Systems Area in Oviedo University (Spain). ACM and IEEE voting member. His research interests include Object-Oriented technology, Language Processors, Human-Computer Interface, Web Engineering, Modeling Software with BPM, DSL and MDA.

Jose Emilio Labra Gayo obtained his PhD. in Computer Science Engineering in 2001 at the University of Oviedo with distinction. Since 2004, he is the Dean of the School of Computer Science Engineering from the University of Oviedo and since 2004, he is also the leader of the WESO research team that belongs to the OOTLab research group. The WESO group (Semantic Web Oviedo) collaborates with the CTIC Foundation in several projects like the development of an Information Services model based on Semantic Web for the Austrian Government or participating in several R&D projects like "Oviedo3, Integral Object-Oriented System", "JEDI (Java Enabled Database Access Over the Internet)", "Components and Objects Interoperability by Means of the Oviedo3 Integral Object-Oriented System", "Applying Static Analysis Technologies to Software Quality", "Proof Carrying Code" and "Development and implementation of a Web service for the analysis and interpretation of financial information in XBRL format." Apart from teaching in the University of Oviedo, he has been regularly invited to teach in several doctorate or postgraduate courses in the Universidad Técnica Federico Santa María, in Chile, and the Universidad Pontificia de Salamanca, in Madrid, Spain. His research interests are Semantic Web technologies, programming languages and Web engineering, where he has published a number of papers in selected conferences and journals. He participates in several committees like the International Conference on Web Engineering, the International Workshop on Social Data on the Web, and the Web Services and Service Oriented Applications Spanish Conference.

Jordán Pascual Espada is a PhD student in the Computer Science Department of the University of Oviedo (Spain). He has a B.Sc. in Computer Science Engineering and a M.Sc. In Web Engineering. His research interests include the the Internet of Things, exploration of new applications and associated human computer interaction issues in ubiquitous computing and emerging technologies, particularly mobile and embedded systems.

Margaret Fitzgerald-Sisk is a graduate student in the Work and Human Resource Education, Department of Organizational Leadership, Policy, and Development, University of Minnesota. She has worked in instructional design, development, and delivery for business and academia for more than 15 years. Her research and writing includes topics on instructional design, faculty development, program redesign, and higher education teaching.

Juan Miguel Gomez-Berbís is an Associate Professor at the Computer Science Department of the Universidad Carlos III de Madrid. He holds a PhD in Computer Science from the Digital Enterprise Research Institute (DERI) at the National University of Ireland, Galway and received his MSc in Telecommunications Engineering from the Universidad Politécnica de Madrid (UPM). He was involved in several EU FP V and VI research projects and was a member of the Semantic Web Services Initiative (SWSI). His research interests include Semantic Web, Semantic Web services, business process modelling, b2b integration and, recently, bioinformatics.

Francesca Grippa, PhD, is an Assistant Professor in eBusiness Management at the Scuola Superiore ISUFI, University of Salento (Lecce, Italy). Her current research interest is in applying social network analysis to business and learning communities. Francesca holds a Ph.D. in e-Business Management and a MA in Business Innovation Leadership from the University of Salento and a BS in Communication Sciences from the University of Siena, Italy. In 2005 and 2006, she was Visiting Scholar at the MIT Sloan Center for Collective Intelligence.

Enrique Herrera-Viedma is Senior Lecturer with the Computer Science and Artificial Intelligence Department and Associate Dean of Academic Management at the University of Granada (Spain). His research interests range from fuzzy logic, decision making systems, information retrieval, and Web quality evaluation to digital libraries. He is author of more than 50 research works published in impact journals, member of the editorial committee of *Fuzzy Set and Systems* and *Soft Computing*, top-cited researcher (Top 1%) in the engineering domain according to the "*Essential Indicators*" database, and has a H=17 index according to the *Web of Science* database.

Luis Iribarne received his BSc degree in computer science from the University of Granada, and the MSc and PhD degrees in computer science from the University of Almería, Spain. From 1991 to 1993, he worked as a lecturer at the University of Granada. In 1993, he collaborated as IT Service Analyst at the University School of Almería, and he served for nine years as a lecturer in the Polytechnic School at the University of Almería. From 1993 to 1999, he worked in several national and international research projects on distributed simulation and geographic information systems (GIS). In 2001 he joined Data, Knowledge and Software Engineering Group and then became Associate Professor, University of Almería in 2002. In 2002, he co-founded the Systems Information Group, and in 2007, he founded the

Applied Computing Group (leader). His research interests include model-driven engineering (MDE), model transformations, ontology-driven engineering (ODE), human-computer intereraction (HCI) for advances on user interfaces, system modelling, component-based software development (CBSD), COTS components, trading, software services, agents & MAS, and UML design. He is an IEEE and EUROMICRO member.

Tania Kerkiri holds a Diploma in Computer Engineering and Informatics from the University of Patras and a Doctor's degree in Semantic Web based e-learning systems (Univ Of Macedonia, Greece). Her international presence by publications on research & applications proves her expertise in: Semantic Web/e-learning & recommendation systems, e-commerce system development, and database administration over Internet applications. She is an Information Technology Consultant, System Administrator & Programmer, Personnel Trainer & Computer Science Tutor at many Companies & Universities. She is Prior-Vice President & now Secretary General of North Hellas Electrical Engineers & Inf. Institute; Greek-Technical-Chamber Member; and a National Accreditation Center certified Tutor.

Dimitris Konetas, Secretary of the Board of Directors of Greek Union of Computer Scientists, graduated from Computer Engineering & Informatics Department of University of Patras, (1992), PhD candidate at ICT-Distance Education Laboratory of University of Ioannina ("Coefficients determining success of blended learning of tele-trainers"). His research interests include: online learning environments, digital libraries, analysis of algorithms, Information Systems, vocational training/guidance, environmental education and sustainability. He worked as: Information Technology consultant, System (Librarian) Administrator in E. C. projects/personnel training programs. Currently he teaches Computer Science in the 4th Vocational Secondary School of Ioannina and Technological Educational Institution of Epirus Dept. Communications, Informatics and Management.

Marco De Maggio, PhD, in "e-business" at eBusiness Management Section of Scuola Superiore ISUFI-University of Salento, Italy. Bachelor Degree in Economics at University of Lecce, and Chartered Accountant. His research field concerns the development of methodologies for the analysis and management of organizational learning patterns inside organizations and communities of practice. His focus is mainly on the development of tools and methodologies for the monitoring of the organizational behavior responsible for social capital creation. Visiting Scholar at MIT–Boston, MA, he experimented the application of content and social network analysis supported by computer-aided systems for the improvement in the analysis of virtual communities.

Oscar Sanjuán Martínez is a Lecturer in the Computer Science Department of the University of Oviedo. Ph.D. from the Pontifical University of Salamanca in Computer Engineering. His research interests include Object-Oriented technology, Web Engineering, Software Agents, Modeling Software with BPM, DSL and MDA.

Rubens Nascimento Melo is a Senior Professor and Researcher in the field of Databases at the Computer Science Department of the Pontifical Catholic University of Rio de Janeiro, PUC-Rio. He holds a B.Sc. in Electronic Engineering (1968), a M.Sc. (1971) and a Ph.D. (1976) degree in Computer Science from the Air Force Institute of Technology (ITA) in Sao Paulo. Currently, he is Associate Professor at PUC-Rio where he leads the Database Research Lab (TecBD). One of his current interests is the application of database technology to distance learning. In this field he has also served as Director of the Centre of Distance Education under the Vice-Rectory for Academic Affairs of PUC-Rio.

Antoni Olivé is a professor of Information Systems at the Universitat Politècnica de Catalunya in Barcelona. He has worked in this field during over 20 years, mainly in the university and research environments. His main interests have been, and are, conceptual modeling, requirements engineering, Information Systems design and databases. He has taught extensively on these topics. He has also conducted research on these topics, which has been published in international journals and conferences. He is a member of IFIP WG8.1 (Design and evaluation of Information Systems), where he served as chairman during 1989–1994.

Nicolas Padilla received his BSc and PhD degrees in computer science from the University of Almeria (Spain), and the MSc degree in computer science from the University of Granada (Spain). From 1991 to 1993, he worked as an Associate Professor at the University of Granada. From 1993 to 2002, he worked as Associate Professor in the Polytechnic School at the University of Almería. From 2002 to 2009, he served as a lecturer and from 2009 to today, he server as a professor, both of them in the same school at the University of Almería. In 2007, he co-founded the Applied Computing Group (co-leader). His research interests include cooperative systems (CSCW), Model-Driven Engineering (MDE), Human-Computer Interaction (HCI), agents & MAS, and UML design.

Giuseppina Passiante is associate professor at the Department of Innovation Engineering and director of the Center for Business Innovation, University of Salento, Italy. Currently, her research fields concern the e-business management, and more specifically the management of learning organizations and learning processes in the net-economy. Her focus is mainly on the development of intellectual capital, both in entrepreneurial and academic organizations. She is also an expert in the development of local systems versus information and communications technologies (ICTs), ICTs and clusters approach, and complexity in economic systems. In these research fields, she has realized programs and projects, and published several papers.

Eduardo Peis is lecturer with the Department of Library and Information Science at the University of Granada (Spain), and one of the first doctors in Information Science in Spain. His research areas include the application of Semantic Web technologies to Web information retrieval enhancement, Information Systems and digital libraries. As a result of his academic and researching activity, he has published several works in some of the main journals in the *Information Science* and *Computer Science* areas, and presented numerous papers in international meetings and conferences.

B. Cristina Pelayo G-Bustelo is a Lecturer in the Computer Science Department of the University of Oviedo. Ph.D. from the University of Oviedo in Computer Engineering. Her research interests include object-Oriented technology,Web Engineering, eGovernment, Modeling Software with BPM, DSL and MDA.

Alejandro Rodríguez-González is a Research Assistant and Ph.D. candidate of the Computer Science Department in University Carlos III of Madrid. He is working as researcher in Computer Science Department involved in several projects of the Spanish ministry of industry. His main research interests are Semantic Web, artificial intelligence, and the building of medical diagnosis systems using these techniques. He has a degree in Computer Science by University of Oviedo and an M.Sc in Computer Science and technology in the specialty of Artificial Intelligence. He is also studying an M.Sc in Engineering Decision Systems at University Rey Juan Carlos.

Sean Wolfgand Matsui Siqueira is an Assistant Professor at the Department of Applied Informatics, Federal University of the State of Rio de Janeiro (UNIRIO), Brazil, where he teaches courses in Databases, Information Systems and Semantic Web. He holds a M.Sc. (1999) and a Ph.D. (2005) in Computer Science, both from the Pontifical Catholic University of Rio de Janeiro (PUC-Rio), Brazil. His research interests include knowledge representation, Semantic Web, ontologies, information integration, semantic models, user models, e-learning, social Web, and music Information Systems. He has experience in the Computer Science area, with focus on Information Systems and technology enhanced learning. He has participated in some international research projects and has published more than 70 papers for conferences, journals, and books.

Jairo Francisco de Souza is an Assistant Professor at the Department of Computer Science, Federal University of the Juiz de Fora (UFJF), Brazil, where he teaches courses in Software Engineering and Semantic Web. He holds a M.Sc. (2007) from Federal University of Rio de Janeiro (UFRJ) and he is a Ph.D. candidate in the Pontifical Catholic University of Rio de Janeiro (PUC-Rio), Brazil, both in Computer Science area. His research interests include knowledge representation, Semantic Web, ontologies, information integration, and semantic models.

Nouha Taifi holds a PhD in eBusiness Management, eBusiness Management Section – Scuola Superiore ISUFI, University of Salento, Italy. She also holds a MSc in Business and Economics, International University of Dalarna, Sweden, and a BBA, University of AlAkhawayn University, Ifrane. Her main research interests are information and knowledge management systems for innovation, organizational structures optimization for new knowledge creation, inter-organizational collaboration in complex product development. In the cPDM laboratory, she was also post-doc researcher in the area of new product development processes in the automotive and aerospace industries.

Robert D. Tennyson is Professor of Educational Psychology at the University of Minnesota. He is editor of a professional journal, *Computers in Human Behavior*. He also serves on editorial boards for four other journals. His research and publications include topics on problem solving, concept learning, intelligent systems, testing and measurement, instructional design, and advanced learning technologies. He has directed NATO sponsored workshops and advanced study institutes on automated instructional design and delivery in Spain and Norway. He has authored over 300 journal articles, books and book chapters.

Salvatore Totaro, Eng., holds a Degree in Electronic Engineering at the Politecnico of BARI (Italy). He is a Research Fellow at the eBusiness Management Section of Scuola Superiore ISUFI - University of

Salento (Lecce, Italy). His areas of interest are knowledge management and artificial intelligence. He is actively involved as software, hardware and net infrastructure designer and developer in several national and international projects. He is teacher of "Data Base design for flexible historical and geographical system" and "Software packages for data base implementation" in international higher education courses.

Avgoustos Tsinakos holds a BSc in Physics (1992), MSc in Applied Artificial Intelligence (1993), an AGDDE(T) Advanced Graduate Diploma in Distance Education Technology (1999) a MDE Master in Distance Education (2000) and Ph.D. degree on "Artificial Intelligence and Internet" (2002). He is an Associate Professor, Department of Industrial Informatics, University of Kavala Institute of Technology and an Adjacent Professor of Athabasca University of Canada. He also acts as Coordinator of Master of Distance Education in Greece, on behalf of Athabasca University. He coordinated 13 research projects, is a member of European Expert Evaluators Team (2006-today) on the field of Distance Education, and participated in more than 32 world/international/national conferences as invited speaker/presenter, holding a number of publications on the area of distance learning, student modeling and artificial intelligence.

Antonio Zilli graduated in physics and is a researcher at the Center for Business Innovation, University of Salento, Italy. His research interests regard technologies that enable powerful virtual collaboration in communities and organizations. He participated in research projects on Semantic Web and on the design of technologies and tools for managing ontology and semantically annotated documents. He is one of the editors of the book 'Semantic Knowledge Management: an Ontology-based Framework' published by IGI Global. He is also senior researcher in the security and privacy issues of IT-based knowledge sharing in value networks.

Index

D

E

F

G

H

I

O

P

Q

R

S

T

U

V

W

Y